Springer Geography

For further volumes:
http://www.springer.com/series/10180

Shinsuke Kato • Kyosuke Hiyama
Editors

Ventilating Cities

Air-flow Criteria for Healthy and Comfortable Urban Living

 Springer

Editors
Shinsuke Kato
Institute of Industrial Science
The University of Tokyo
4-6-1, Komaba, Meguro-ku
Tokyo 1538505, Japan
kato@iis.u-tokyo.ac.jp

Kyosuke Hiyama
Institute of Industrial Science
The University of Tokyo
4-6-1, Komaba, Meguro-ku
Tokyo 1538505, Japan
hiyama@iis.u-tokyo.ac.jp

ISBN 978-94-007-2770-0   e-ISBN 978-94-007-2771-7
DOI 10.1007/978-94-007-2771-7
Springer Dordrecht Heidelberg London New York

Library of Congress Control Number: 2011943356

Cover design: © Spectral-Design – Fotolia.com

Printed on acid-free paper

Springer is part of Springer Science+Business Media (www.springer.com)

Preface

Most of the population on the earth lives in urban areas. Blocks of buildings in urban areas form windbreaks and reduce wind speeds compared with bare regions. Thus, most of the people in the world live in environments with artificially weakened wind. In their living environments, anthropogenic heat and contaminant is also generated with certain extent by human activities. During the summer, dense urban areas thereby suffer from the urban heat island phenomenon, an urban climate problem.

Wind is a stochastic phenomenon that is mainly driven by atmospheric Rossby waves. The direction of wind varies with passing high or low atmospheric pressure fronts, and the wind stream is not steady, with up and down streams changing frequently. Few books consider the environmental concerns related to wind, especially concerning the weakened wind in urban areas. It is somewhat difficult for civil engineers and civil engineering students studying urban built environments to comprehend the characteristics of urban wind, especially its ventilating characteristics. This book provides the latest knowledge related to urban wind at the pedestrian height from the ground in details.

To create new integrated knowledge for sustainable urban regeneration, the Center for Sustainable Urban Regeneration (cSUR), The University of Tokyo, was established. The center coordinates international research alliances and collaboratively engages with common issues of sustainable urban regeneration. This book presents one of the achievements of the new integrated approach toward sustainable urban regeneration.

Tokyo, Japan   Shinsuke Kato

Contents

Contributors

Zhen Bu Mott MacDonald (Shanghai, China), Unit 2601, 398 Caoxi Bei Road, Xuhui District, Shanghai, 200030, China

Hiroshi Hayami Environmental Science Research Laboratory, Central Research Institute of Electric Power Industry, 1646 Abiko, Abiko-shi, Chiba 2701194, Japan

Kyosuke Hiyama Institute of Industrial Science, The University of Tokyo, 4-6-1, Komaba, Meguro-ku, Tokyo 1538505, Japan

Tomomi Hoshiko Graduate School of Engineering, The University of Tokyo, 7-3-1 Hongo, Bunkyo-ku, Tokyo 1138656, Japan

Hong Huang Center for Public Safety Research, Department of Engineering Physics, Tsinghua University, Beijing 100084, China

Shinsuke Kato Institute of Industrial Science, The University of Tokyo, 4-6-1, Komaba, Meguro-ku, Tokyo 1538505, Japan

Yoichi Kawamoto School of Engineering, The University of Tokyo, 7-3-1, Hongo, Bunkyo-ku Tokyo 1138656, Japan

Mai V. Khiem Applied Climate Division, Vietnam Institute of Meteorology, Hydrology and Environment, 23/62 Nguyen Chi Thanh, Ha Noi, Viet Nam

Mahmoud Farghaly Bady Mohammed Faculty of Engineering, Assiut University, Assiut 271516, Egypt

Fumiyuki Nakajima Environmental Science Center, The University of Tokyo, 7-3-1 Hongo, Bunkyo-ku, Tokyo 1110033, Japan

Keisuke Nakao Faculty of Engineering, The University of Tokyo, 4-6-1, Komaba, Meguro-ku, Tokyo 1538505, Japan

Ryozo Ooka Institute of Industrial Science, The University of Tokyo, 4-6-1, Komaba, Meguro-ku Tokyo 1538505, Japan

Tassanee Prueksasit Faculty of Science, Chulalongkorn University, Phayathai Road, Pathumwan, Bangkok 10330, Thailand

Hom Bahadur Rijal Department of Environmental & Information Studies, Tokyo City University, 3-3-1 Ushikubo-nishi, Tsuzuki-ku, Yokohama 224-8551, Japan

Kazuo Yamamoto Environmental Science Center, The University of Tokyo, 7-3-1 Hongo, Bunkyo-ku, Tokyo 1110033, Japan

Hiroshi Yoshikado Graduate School of Science and Engineering, Saitama University, 255, Shimo-Okubo, Sakura-ku Saitama-shi Saitama 3388570, Japan

Chapter 1
Introduction

Shinsuke Kato and Kyosuke Hiyama

Keywords Wind environment • Low wind • Urban environment • Criteria • Ventilation performance

1.1 Measurement and Evaluation of the Ventilation Through and Over Urban Blocks

What is the ideal wind environment in urban areas for people to live healthy and comfortable lives? What is required to realize such wind environments in urban areas? These simple questions cannot be answered easily.

Many factors that determine the wind environment in urban areas are natural phenomena that are beyond the power of human intervention. That is, they cannot be controlled by humans easily. When considering the safety of constructions against strong wind, the possibility of controlling the wind strength is never considered. At most, the structures are designed with a factor of safety from a stochastic point of view by predicting the strength of the wind stochastically. In other words, the countermeasures taken against strong wind are extremely passive. However, the circumstances change when the wind is sufficiently strong not only to damage the constructions but also to affect daily living. The wind strength is controlled to an extent by windbreaks and/or windbreak fences to provide a wind environment acceptable to pedestrians and residents. The building-induced winds that occur due to high buildings should be controlled by the building shape and/or windbreak fences to a status acceptable to pedestrians and residents.

S. Kato (✉) • K. Hiyama
Institute of Industrial Science, The University of Tokyo, 4-6-1, Komaba,
Meguro-ku, Tokyo 1538505, Japan
e-mail: kato@iis.u-tokyo.ac.jp; hiyama@iis.u-tokyo.ac.jp

S. Kato and K. Hiyama (eds.), *Ventilating Cities: Air-flow Criteria for Healthy and Comfortable Urban Living*, Springer Geography, DOI 10.1007/978-94-007-2771-7_1, © Springer Science+Business Media B.V. 2012

Weaker wind is also a concern. The wind in urban areas is expected to dilute pollutants and transport them away from the area and exhaust heat produced within the city so that these factors do not affect daily living. The wind has a corresponding body-cooling ability if the ambient temperature is lower than the body temperature. It effectively cools the human body and buildings that receive heat from solar radiation. What is the status of wind required to protect daily living? What kind of device is needed to ensure wind that is effective to protect people from pollutants and exhaust heat? What can prevent the benefits of useful wind? The researchers in wind engineering continue to discuss these questions in terms of "low wind environments in urban areas." However, because the damage and disadvantages of insufficient wind are not as clear as wind damage caused by strong winds, this topic has been treated lightly. The countermeasure to low wind seems to not have an answer that has achieved social consensus.

The concentration of buildings in urban areas degrades ventilation as well as the effects of dilution/transportation of pollutants and heat exhaust. Currently, the building coverage and floor area ratio, which indicate building density, are restricted to a limited amount to avoid extreme concentration. Open areas such as parks and roads are also used to control the building density. As a result, the degradation of ventilation in urban areas is prevented to a certain degree. However, the relationship between the building density and ventilation level and the influence of poor ventilation that affects daily living are not sufficiently clear as to be explained quantitatively.

In this book, the argument will be developed with those questions and problems as starting points. First, the following information will be provided through a case study: the contribution of a low wind environment to improve the living environment in urban areas, the risk in living environments when a low wind environment is lost, and the measurement and concept of a minimum standard regarding the low wind environment required to provide healthy and comfortable life in the city. To discuss this low wind environment, evaluation methods are required to quantitatively explain the building density and the relationship between urban geometry and ventilation level. Therefore, disregarding the influence of regionality, a globally applicable evaluation method will be proposed.

1.2 Outline of Chapters

This book is divided into two sections.

Part I is entitled "Wind Environment and Urban Environment." It consists of four chapters, Chaps. 2, 3, 4, and 5. Here, the benefit of a low wind environment, such as the effect of sea breeze in reducing the heat island phenomenon, and the improvement in the thermal comfort of the human body with cool breeze will be introduced. This chapter also introduces the possible risk when the low wind environment is insufficient, a health damage risk assessment due to the transport of exhaust emissions, and the evaluation of ventilation performance against a sudden diffusion of hazardous material, such as biological or chemical terrorism. They will be explained in detail through a case study performed by computer simulation and/or survey investigation.

Part II is entitled "Criteria for Assessing Breeze Environment." It consists of two chapters, Chaps. 6 and 7. This part introduces the concept required to realize the wind environment, which enables a healthy and comfortable life. To secure healthy and comfortable living conditions, it is important to define the minimum wind conditions required. Then the regulation required to include this wind environment in city planning must be discussed. In this case, "specific safety guidelines" that make a certain level of performance obligatory as performance target are considered important. A method to quantitatively evaluate the wind environment is required when defining specific safety guidelines. This part includes the authors' ideas regarding the definition of a wind environment to secure healthy and comfortable living conditions, the quantitative evaluation method of such a wind environment, and the regulation concept to reflect the wind environment in city planning.

1.2.1   Part I: Wind Environment and Urban Environment

Part I Chap. 2 is entitled "Sea Breeze Blowing into Urban Areas: Mitigation of the Urban Heat Island Phenomenon." In recent years, various sorts of degradation in city environments have been discussed due to sudden urbanization. One of the problems that have gained attention is "heat islands," which means that the ambient temperature in the urban area is increased compared to the surrounding rural areas. As a solution to this problem, a countermeasure to take in the cold air generated in green land and waterscapes within or close to the city has been suggested. Among them, the utilization of the sea is particularly promising because it can generate a large amount of cold air. Maritime transportation plays a huge role in city development. Therefore, many of the large cities in the world are located in areas next to large bodies of water. About half of the global population lives within 100 km of a coastline, which is another reason that using sea breeze to cool cities is expected to have a great benefit. In this chapter, a case study intended for Tokyo, which was chosen among megacities that show the heat island phenomenon, will be introduced. Using computer simulations, the effect of sea breeze was quantitatively evaluated by considering the influence of city planning, including land use, on the sea breeze utilization. According to this case study, the change in land use between 1976 and 1997 caused a temperature rise of 0.86°C in the city area. It also delayed the sea breeze arrival to an inland area (about 30 km from the coastline) by approximately 20 min. The thoughts regarding the countermeasures for the heat island phenomenon and the evaluation method of the cooling effect in the city area using wind will be introduced through this case study.

Part I Chap. 3 is entitled "Thermal Adaptation Outdoors and the Effect of Wind on Thermal Comfort." Temperature, humidity, radiation, and airflow, collectively, the wind environment, are listed as external factors that affect the warm-cold sense of the human body. In contrast to the indoors, where the environment can be controlled, these factors are influenced greatly by changes in the natural environment outdoors. On the subject of indoor environments, the thermal comfort index to determine the goal value of air-conditioning is proposed with the assumption that the indoors can be controlled to an ideal thermal environment. On the other hand, outdoors, the main focus is on revealing how humans feel under variable natural environments. However, because of the adaptability of humans to the surrounding environment, the amenity level varies with region. This chapter explains the research on outdoor amenity based on the environmental adaptability of humans and regionality. In addition, information on the concept of an outdoor thermal environment in each region will be provided. The outdoor thermal environment is not uniform. It is influenced by region, especially the average temperature in each season. For example, a temperature lower than the ideal value determined by examining thermal balance is regarded as comfortable in cold areas, but a temperature higher than the ideal value is regarded as comfortable in hot areas. The wind environment is also influenced strongly by regionality. High wind velocity causes discomfort in cold areas, but in hot areas, it improves comfort. The outdoor wind environment depends on the city shape. Therefore, the wind environment is the only factor that is controllable to some extent in natural environments. This chapter gives important knowledge when examining the wind environment of a city and the amenity of outdoor environments.

Part I Chap. 4 is entitled "Health Risk of Exposure to Vehicular Emissions in Wind-Stagnant Street Canyons." The air pollution problem has existed throughout the ages. For example, the air pollution caused by industrial exhaust has led to serious harm, and therefore, many studies were performed. As a result, various data were obtained regarding risk evaluation. However, the air pollution problem has been diversified in recent years. First, the transportation mechanism of exhaust from chimneys, such as industrial exhaust, is different than that of pollutants generated near the ground. When the pollutant occurs near the ground, the transportation is obstructed by the surrounding structures. The pollutant becomes stagnant in the street canyon and increases the health risk. With an insufficient wind environment, the risk can become extremely high. Therefore, the consideration of the wind environment and city characteristics is very important when evaluating risk. Understanding the health risk, including the reaction speed of each material, is an important issue because the risk evaluation target materials are also diversifying. This chapter introduces the risk evaluation result of PAHs based on in situ measurements. Among the automobile air pollutant problems intensifying in Asia, PAHs have an insufficient health risk evaluation. Through this case study, the idea of risk assessment in urban areas will be explained.

Part I Chap. 5 is entitled "Pollutant Transport in Dense Urban Areas." To prevent health damage due to air pollution, the first step is to identify the source of a pollutant and reduce its generation. The conventional air pollution countermeasures generally follow this idea because the air pollution and health damage due to industrial exhaust appeared with industrialization. However, in recent years, risk management not only for the industrial air pollution but also for the sudden pollution, such as gas tank leaks or diffusion of hazardous material due to BC terrorism, is required. These accidental, sudden emissions are difficult to control with the conventional method. Therefore, as a second best solution, the method to send fresh air to the source of pollution to dilute it and "ventilate" it appropriately is considered to be effective. In this case, it is required to accurately predict whether wind will transport and dilute the pollution. When considering the pollution from industrial exhaust, the pollutant exits from the chimney projecting from the city block. Hence, the building density has no effect on diffusion. The transportation and diffusion simply depend on the ambient air. In addition, because an examination of the long-term exposure is required when evaluating health risk, the consideration of the average value was sufficient for pollutant concentration. However, when the pollutant occurs within the urban area, such as with BC terrorism and automobile exhaust gas, the diffusion is influenced greatly by the building shapes. The increase in the number of voids, where the air become stagnant and so does the pollutant, in the urban area due to rapid urbanization also complicates the pollutant diffusion problem. To consider the health risk of short-term exposure, it must be determined whether the concentration exceeded the threshold amount, even for an instant, that causes a health problem by predicting the concentration fluctuation. In this chapter, the conventional pollutant diffusion model targeting industrial exhaust will be explained first. In addition, the information including the characteristics of the concentration fluctuation regarding the pollutant diffusion in the actual city will be given. The information was obtained through a case study performed by wind tunnel tests using the model of actual urban areas (Tokyo). When the source of the pollutant and the observation point were close, a high concentration material transportation was observed through the path created by the buildings. In addition, the properties of diffusion, such as the occurrence of a vortex due to the buildings, will be explained. A vortex plays a huge role in pollutant diffusion in the vertical direction. It is clearly stated that the difference in the positional relationship between the occurrence point and observation point (e.g., the distance or urban shape of the transportation path) is a characteristic of its concentration fluctuation. The information was given to predict the diffusion.

1.2.2 Part II: Criteria for Assessing Breeze Environment

Part II Chap. 6 is entitled "Legal Regulations for Urban Ventilation." People wish to have a healthy and comfortable life, and they have a right to do so. However, when everyone insists on their maximum rights, the right of their neighbors can be violated. This insistence of rights can often escalate into a conflict. To solve this problem and secure the best living environment for the whole urban area, it is important to protect everyone's rights with appropriate legal regulations.

This chapter gives an overview of existing legal regulations regarding the wind environment necessary for a healthy and comfortable life. However, an existing countermeasure against air pollution focuses on the source control, not on the examination of the city shape to promote ventilation. Hence, legal regulation that is approved based on sufficient examination of the relationship between the city shape and ventilation performance does not exist at the present moment. Therefore, based on the method of examining the city shape by defining specific safety guidelines, the authors reviewed the ideas of precursors regarding ventilation and developed the concept of legal regulation to secure better ventilation.

Part II Chap. 7 is entitled "New Criteria for Assessing the Local Wind Environment at the Pedestrian Level and the Applications." This chapter proposes the concept of the minimum standard and the measurement method of ventilation required for a healthy and comfortable life:

1. Ventilation requires continuous airflow. Therefore, it is rational to evaluate it in the continuous region (the simplest is a spatial average) and not as a value at a point.
2. Because the wind is a stochastic phenomenon, the ventilation should be evaluated as a statistical value over a certain period of time, such as spring, summer, fall, and winter or yearly. Determining the probability of days that fulfill the required wind environment is better for human recognition than evaluating the average rates of a period.
3. Kinetic energy is used an index to improve thermal comfort because a scalar quantity is more convenient in indicating the strength of wind than a vector quantity.
4. The kinetic energy that indicates wind strength does not directly express the ability to discharge pollutants or exhaust heat. Therefore, the ventilation efficiency index, which indicates the transportation efficiency of pollutants and exhaust heat by wind, is used as an index to express discharging ability.

Based on these concepts, the quantification of an average kinetic energy in the target area for comfort is proposed. For health (safety), the quantification of ventilation performance using the ventilation efficiency index is also proposed. This minimum standard value for kinetic energy is 0.05 m^2/s^2, and value for a purging flow rate (PFR) is above 60 times/h. This kinetic energy is defined by the kinetic energy of the wind that people acknowledge as cool (0.3 m/s). The PFR is 10 times faster than that required to rapidly clear the pollutants when they occur indoor (6 times/h). By clearing the tenfold amount, it is thought that clean ambient air is obtained at all times, even when rapidly ventilating indoors. Then a stochastic expression using these minimum standard values is proposed. For example, the following methods are listed as options:

1. There is more than 1 day in a week where the probability that the kinetic energy of wind has a cooling effect on the human body, meaning that the kinetic energy exceeds 0.05 m^2/s^2.

2. There are more than 6 days in a week where the probability that safety can be expected because wind reduces the health risk, meaning that PFR is 60 times/h in a week.

This chapter contains an example to evaluate this method in detail through the case study using street canyons and a concentrated city. An enrichment of the understanding of the evaluation method and its procedure is the aim of this chapter.

Part I
Wind Environment
and Urban Environment

Chapter 2
Sea Breeze Blowing into Urban Areas: Mitigation of the Urban Heat Island Phenomenon

Yoichi Kawamoto, Hiroshi Yoshikado, Ryozo Ooka, Hiroshi Hayami, Hong Huang, and Mai V. Khiem

Abstract Currently, about 50% of the world's population is living in urban areas, and that figure is predicted to continue to increase (United Nations, Department of Economic and Social Affairs Population Division, Population Estimates and Projections Section (2009) World urbanization prospects: the 2009 revision). On the other hand, many cities are facing problems caused by urbanization. The urban heat island phenomenon, one of the urban climate problems, is a typical

Y. Kawamoto (✉)
School of Engineering, The University of Tokyo, 7-3-1, Hongo,
Bunkyo-ku Tokyo 1138656, Japan
e-mail: kwmt@iis.u-tokyo.ac.jp

H. Yoshikado
Graduate School of Science and Engineering, Saitama University, 255, Shimo-Okubo,
Sakura-ku Saitama-shi Saitama 3388570, Japan
e-mail: yosikado@mail.saitama-u.ac.jp

R. Ooka
Institute of Industrial Science, The University of Tokyo, 4-6-1, Komaba,
Meguro-ku Tokyo 1538505, Japan
e-mail: ooka@iis.u-tokyo.ac.jp

H. Hayami
Environmental Science Research Laboratory, Central Research Institute of Electric
Power Industry, 1646 Abiko, Abiko-shi, Chiba 2701194, Japan
e-mail: haya@criepi.denken.or.jp

H. Huang
Center for Public Safety Research, Department of Engineering Physics,
Tsinghua University, Beijing 100084, China
e-mail: hhong@tsinghua.edu.cn

M.V. Khiem
Applied Climate Division, Vietnam Institute of Meteorology, Hydrology and Environment,
23/62 Nguyen Chi Thanh, Ha Noi, Viet Nam
e-mail: khiem@vkttv.edu.vn

S. Kato and K. Hiyama (eds.), *Ventilating Cities: Air-flow Criteria
for Healthy and Comfortable Urban Living*, Springer Geography,
DOI 10.1007/978-94-007-2771-7_2, © Springer Science+Business Media B.V. 2012

environmental problem encountered in dense urban areas in summer. The use of the sea breeze to mitigate the urban heat island phenomenon has attracted attention in coastal cities. Some statistics show that about 40% of the world's population lives within 100 km of the coast (World Resources Institute, Fisheries (2007) Population within 100 km of coast). Further investigation of the environment in the urban area near the coast is, therefore, important. In this chapter, Tokyo is targeted for investigation. Tokyo is the Japanese capital, and its surrounding region, the Tokyo metropolitan area, comprises a circular area with a radius of 50 km and a population of over 30 million. Within this area, the sea breeze from Tokyo Bay is an important factor mitigating the air temperature rise in summer. However, ongoing urbanization could be changing not only the mechanism of the energy balance on the urban surface but also the sea breeze system in the region. To clarify the effects of urbanization, a mesoscale meteorological model was adopted for analysis. Simulation results suggest that expansion of the Tokyo metropolitan area from the 1970s to the 1990s has induced a temperature rise near the ground and that the difference is largest in inland areas. Moreover, the time of sea breeze penetration is delayed in suburban areas. These results suggest that the ongoing urbanization process could raise the air temperature and change the sea breeze system in the Tokyo metropolitan area.

Keywords Urban heat island phenomenon • Urban climate • Urban boundary layer • Mesoscale meteorological model • Land use change

2.1 Overview of Urban Heat Island Phenomenon in the Tokyo Metropolitan Area

2.1.1 Overview of Research on Urban Climate and the Urban Heat Island Phenomenon

In the twentieth century, rapid urbanization produced many environmental problems throughout the planet, including air pollution, water pollution, ground pollution, environmental noise, etc. The urban climate problem is a characteristic environmental effect caused by dense urban areas. Above all, the urban heat island (UHI) phenomenon is a typical urban climate problem found in large cities.

The UHI phenomenon was first recognized in the nineteenth century. The earliest urban climate research was conducted by Luke Howard (1772–1864). Howard was an amateur meteorologist and well known for his nomenclature system of clouds. Furthermore, he made an important contribution to the field of urban climate research by being the first person to detect the urban heat island phenomenon and suggest its causes. From 1806 to 1830, Howard measured various meteorological elements, i.e., wind direction, pressure, temperature, precipitation, and so on, outside London. Comparing these recorded data with measurements taken in the center of London by the Royal Society, Howard found that the mean temperature of the climate was 2.00°F (about 1.1°C) higher in the denser parts of the metropolis compared with the suburban part—with the temperature difference being especially great on winter nights. He summarized that these temperature differences were caused by the effect of anthropogenic factors (i.e., the crowded population and the consumption of great quantities of fuel in fires).

His achievement was published in three volumes titled "The Climate of London" in 1833 (2nd edition). As the original version of "The Climate of London" is now very rare and not normally available for handling, it was republished in 2007 by the International Association for Urban Climate (IAUC) (Howard 2007) in recognition of its importance to the field of urban climatology. It can now be easily obtained and used as a guidebook for urban climatology field.

Since the dawn of urban climatology field and the work carried out by Howard, many kinds of research studies have been conducted. Sundborg made an isotherm map of the urban area of Uppsala, Sweden (Sundborg 1950). A remarkable aspect of this study was that these air temperatures were measured using thermometers mounted on the roof of an automobile. In the late 1960s, Myrup adapted a numerical energy budget model to analyze UHI (Myrup 1969). Landsberg showed the effect of changing surface characteristics in urban areas on UHI through the use of microscale observations (Landsberg and Maisel 1972). Oke, the first president of the IAUC, also conducted various research studies in the field of urban climatology. He compared typical temporal variations in air temperature in urban areas and rural areas and found that the heat island intensity, ΔT_{u-r}, can be defined by the air temperature difference between urban and rural areas, under clear skies, and at a low wind velocity (Oke 1987).

In Japan, urban climate research was carried out from the 1930s onwards in the field of geography. For example, Eiichiro Fukui, the pioneer of climatology in Japan, and his colleagues published their observations mainly in the Association of Japanese Geographers and the Meteorological Society of Japan (Fukuoka and Nakagawa 2010). From the 1970s, the range of urban climate research widened with the increasing incidence of problems caused by air pollution and the rise of air temperature in the city in summer. These researches were carried out in fields that included architecture, civil engineering, atmospheric chemistry, in addition to geography and climatology. Toshio Ojima, former president of the Architectural Institute of Japan, and his colleagues have been continuously recording observations, seeking to clarify and mitigate UHI in Tokyo since the 1970s. Shuzo Murakami, also a former president of the Architectural Institute of Japan, and his colleagues have applied computational fluid dynamics simulation to UHI analysis (Mochida et al. 1997). In the atmospheric chemistry field, to clarify the transportation of atmospheric pollutants affected by the sea breeze from Tokyo Bay, large-scale upper-air observations have been carried out in the Tokyo metropolitan area using balloons by Yoshikado and Kondo (1989).

In recent years, urban climate phenomena, including UHI, and their effects have been investigated in various fields. In addition to the examples noted above, many experimental research studies have also been conducted in the medical and physiological fields because the UHI phenomenon can induce heat stroke in summer.

Furthermore, various methods are now being adopted for both observation and analysis. For example, remote sensing techniques via satellite and airplanes have enhanced observation availability—now no longer limited to in situ observation. For example, radiative temperature observations can be used to illustrate the surface heat island condition, the reflection rate from a land surface (including urban areas) can be utilized in a numerical model to set analysis conditions, and land surfaces and building geometries can also be utilized in numerical model configurations or correlation models examining the relationship between urban geometry and UHI. A number of numerical models (ranging from micro- to macroscale) have been developed and improved upon and are now widely used. Moreover, rapid increases in computational power have made numerical analysis more useful, with analysis domains becoming larger and mesh resolutions becoming finer. As a result, more realistic (but also more complicated) assumption modeling is now becoming available instead of just the simple assumption modeling available previously.

2.1.2   Present State of Urban Heat Island
in the Tokyo Metropolitan Area

Statistics provided by the United Nations Population Division show that about 50% of the world's population is living in urban areas and that this figure is predicted to continue to increase (United Nations 2009). Other statistics show that about 40% of the world's population lives within 100 km of the coast (World Resources Institute 2007). Further investigation of the environment in the urban area near the coast is, therefore, important. In this chapter, Tokyo is targeted for analysis of UHI as one example of a coastal urban area. Tokyo is the Japanese capital, and its surrounding region, the Tokyo metropolitan area, comprises a circular area with a radius of 50 km and a population of over 30 million.

Since the latter part of the twentieth century, the UHI phenomenon in summer has become an increasingly important issue, although the UHI intensity on winter nights is also especially large. The problems caused by UHI in summer can be summarized as follows: (1) changes in the ecosystems of animals and plants, (2) human health problems such as heat stroke and the spread of infectious diseases (while the ozone concentration at ground level also increases because higher temperatures accelerate chemical reactions), (3) increased energy consumption for air conditioning, and (4) the increased possibility of heavy rain in the urban area.

The UHI effect is usually attributed to the following two factors: (1) land cover change and (2) an increase in anthropogenic heat release. As part of the urbanizing process, areas of vegetation have been replaced by building sites (i.e., the pervious surface area has been reduced). As a result of urbanization, the energy balance at the surface in urban areas tends to show a decrease in latent heat flux and an increase in sensible heat flux. The other key factor contributing to the UHI phenomenon is the anthropogenic heat release from buildings, industries, and transportation—directly heating up the atmosphere.

Attention is often focused on the ventilation path in the urban area as a potential mitigation method. The original concept of the urban ventilation path was first implemented in Stuttgart, Germany. Stuttgart is the capital of the state of Baden-Württemberg, and the city center is at the bottom of a basin-shaped valley. In the 1980s, in order to mitigate the air pollution problem which was becoming a serious issue in winter, the ventilation path concept (Luftleitbahne in German) was proposed. The objective of the ventilation path concept in Stuttgart was to maximize the effect of mountain breezes flowing into the city center by means of urban planning—creating green belts, regulating building shapes, and so on. In this case, it is the fresh air from the mountainside that is expected to dilute the air pollutants.

On the other hand, in the Tokyo metropolitan area, it is the sea breeze flowing from coastal areas into inland urban areas that is expected to play an important role in mitigating UHI in summer. The sea breeze is driven by the temperature difference between the air over the sea surface and the land surface. Thus, Tokyo Bay acts as an inlet to the Tokyo metropolitan area for cool, fresh air. Aiming to utilize the sea breeze from Tokyo Bay, a number of research studies, involving both observation and simulation, have been conducted on urban ventilation paths. For example, targeting the large-scale mechanism of the sea breeze coming from Tokyo Bay, Ooka et al. utilized a mesoscale meteorological model to evaluate the mean kinetic energy of the sea breeze over the Kanto Plain in Japan (Ooka et al. 2008). On the other hand, Narita observed sea breeze penetration along a small river in the dense urban area of Tokyo and used this to clarify the microscale phenomena of sea breezes (Narita 2006).

In this chapter, the effect of land cover change on the UHI phenomenon in the Kanto Plain is examined. Ongoing urbanization in the Tokyo metropolitan area could be changing not only the mechanism of energy balance but also the sea breeze system. To clarify the effect of land cover change, a mesoscale meteorological model was adopted and used for analysis.

2.2 Outline of UHI Analysis Using Numerical Simulation

2.2.1 Application of a Mesoscale Meteorological Model to Urban Climate Analysis

First, the definitions used for scale need to be explained. The term "meso" means intermediate between "macro" and "micro," and in the meteorological field, these spatial scales are generally subdivided as follows: microscale (about 2 km or less), mesoscale (from a few kilometers to several hundred kilometers), and macroscale (thousands of kilometers). In some cases, larger scales (above mesoscale) can be divided into synoptic scale (in the order of a thousand kilometers) and global scale

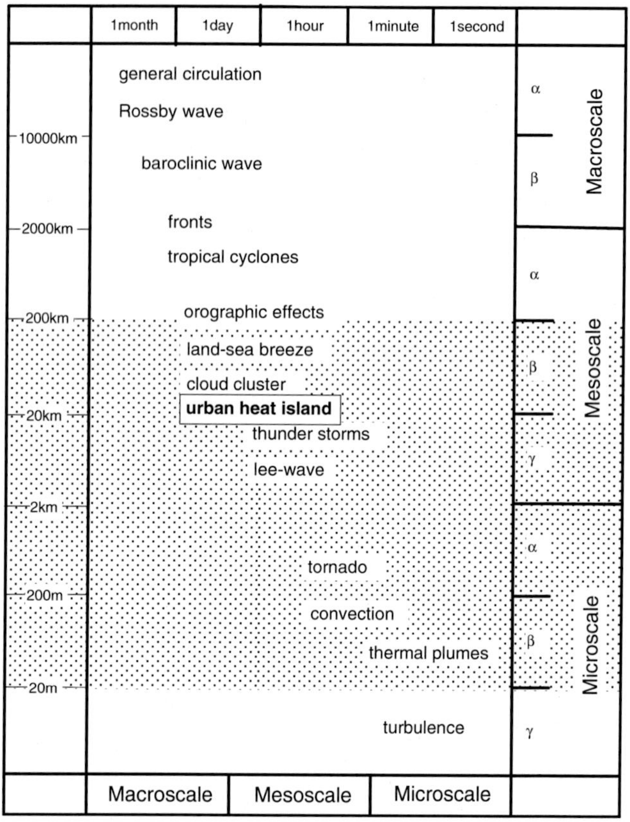

	1month	1day	1hour	1minute	1second		
	general circulation					α	Macroscale
10000km	Rossby wave						
	baroclinic wave					β	
2000km	fronts						
	tropical cyclones					α	
200km	orographic effects						Mesoscale
	land-sea breeze					β	
	cloud cluster						
20km	urban heat island						
	thunder storms					γ	
	lee-wave						
2km							
	tornado					α	Microscale
200m	convection						
	thermal plumes					β	
20m							
	turbulence					γ	
	Macroscale		Mesoscale		Microscale		

Fig. 2.1 A spatial and temporal scale diagram for atmospheric motions (Modified from Gross 1992)

(over the whole planet). According to Orlanski (1975), the mesoscale can be further subdivided into the following subclasses: meso-gamma (2–20 km), meso-beta (20–200 km), and meso-alpha (200–2,000 km). Figure 2.1 shows a spatial and temporal scale diagram for atmospheric motions (modified from Gross (1992)). For example, a "tornado" has a scale of several hundred meters and can therefore be categorized into the microscale in spatial terms. However, it also has a scale of several tens of minutes and can therefore be categorized into the mesoscale in temporal terms. The UHI phenomenon has both a micro-gamma scale (complex terrain-following flow affected by urban structure) and a meso-beta scale (land and sea breeze system) in spatial terms.

To analyze UHI in terms of mesoscale phenomena, e.g., motion of the atmosphere, heat transfer between land surface and atmosphere, and solar radiation and infrared radiation, mesoscale meteorological models have been adopted. In one example, Mochida, Murakami et al., used a mesoscale model to analyze UHI in the Tokyo metropolitan area (Mochida et al. 1997).

Here, the difference between meteorology and climatology has to be noted. The UHI phenomenon is characterized by dense urban areas, which is why it is called an urban climate problem. The term "climate" refers to the trends shown by meteorological phenomena within a particular area. However, the term "meteorological phenomena" refers to atmospheric events, especially the weather and weather forecasting. When analyzing the urban climate, long-term analysis is also preferable. However, long-term climate analysis with a fine resolution mesh, e.g., over a 10-year period with resolution of 1 km for the Tokyo metropolitan area, imposes an extremely high computational load. Furthermore, within the targeted period, the urbanization process will have probably progressed. Land use change associated with the urbanization process is one of the key factors affecting UHI. To investigate how land use change affects UHI, three short-term event analyses were conducted in this study. The targeted event was wide area observation on a summer day in 1986 (Yoshikado and Kondo 1989). On that day, inland sea breeze penetration from Tokyo Bay over the Tokyo metropolitan area was clearly observed. In the control analysis, a representation of this sea breeze penetration was prepared. In two additional cases, land surface parameters were modified to reflect the urbanization process in the Tokyo metropolitan area. An outline of this case study is presented in Sect. 2.2.3.

2.2.2   Outline of Mesoscale Meteorological Model MM5

Simulation models which can be categorized as mesoscale meteorological models possess a large variety of features. Some models use in-house code, and others use open source code; some are hydrostatic models, and others are nonhydrostatic models; some are general-purpose models, and others are specialized for use with a specific meteorological phenomenon; some are operational models used for weather forecasting, and others are used for research purposes. Of these, the open source (or free software) and nonhydrostatic models have been most widely used for research in recent years. The former feature means there are a number of users and that they form a users and developers' community. Therefore, when you encounter a problem, the community may be able to help solve it. Furthermore, you can modify the model by yourself or in collaboration with the community. The latter feature means that the model is capable of high-resolution analysis (e.g., with mesh resolution of several kilometers). Conventional hydrostatic models omit vertical acceleration of the atmosphere. That is to say, the size of the vortex is small compared with the analysis mesh size, and therefore, the amount of vertical momentum averaged over each mesh is negligible. However, as mesh resolution becomes finer, upward and downward motions have to be solved explicitly. The following models possess these features and are often used in urban climatology field: (1) RAMS (the Regional Atmosphere Modeling System) developed by Colorado University (Pielke et al. 1992), (2) MM5 (the fifth-generation Penn State/NCAR Mesoscale Model) developed by NCAR (National Center for Atmospheric Research)

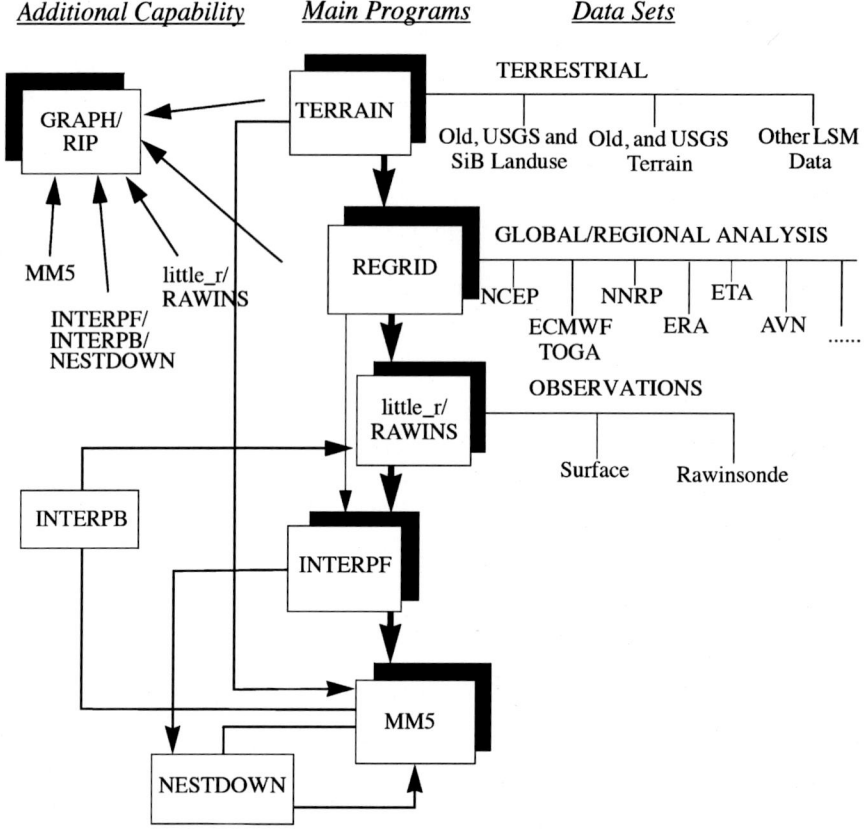

Fig. 2.2 The MM5 modeling system flow chart (Dudhia et al. 2005)

and Pennsylvania State University (Grell et al. 1995), (3) the WRF (Weather Forecast and Research) model (the next-generation model after MM5) developed by NCAR (Skamarock et al. 2008), and (4) NHM (Nonhydrostatic Model) developed by the Japan Meteorological Agency (Saito et al. 2007). To analyze atmospheric conditions throughout the Tokyo metropolitan area, the MM5 model was used in this study. This MM5 model is based on the original version described by Anthes and Warner (1978) and has been modified to broaden its application. For instance, it now features (1) multiple-nest capability, (2) nonhydrostatic dynamics, (3) four-dimensional data assimilation capability, and (4) an increased number of physics options. Figure 2.2 shows the flow chart for MM5 analysis.

The "TERRAIN" program defines the model analysis domain and interpolates 2-D terrestrial data (e.g., land use, elevation, the land-sea mask) from a latitude-longitude global dataset to the analysis domain. The "REGRID" program interpolates the 3-D isobaric meteorological dataset from the latitude-longitude dataset to the analysis domain. These meteorological datasets are used to initialize and set boundary conditions for the analysis. The "INTERPF" program performs vertical interpolation from the pressure level meteorological dataset to the terrain-following σ-coordinate of the MM5 analysis domain. "MM5" is the main program of the MM5 modeling system. It performs time integration using input data from the preprocessing programs. The "GRAPH" and "RIP" programs are postprocessing programs used to visualize the MM5 output. The flow chart shown above illustrates the standard process of MM5 analysis. In addition, the optional programs "LIT-TLE_R" or "RAWINS" are used to interpolate more detailed datasets derived from observation. The programs "INTERPB" and "NESTDOWN" are also optional. "INTERPB" is used to interpolate MM5 output data to pressure level data to be used in "LITTLE_R" or "RAWINS". "NESTDOWN" is used to interpolate from MM5 output data to finer analysis domain input data when a one-way nesting method is adopted. The basic equations used in MM5 are shown in Appendix.

The MM5 system has been modified by the authors to represent different land use fractions in each analysis grid (Kawamoto and Ooka 2008). To reflect the fraction of land use in each analysis mesh, the surface parameters were area-weighted and averaged, as shown in Eq. 2.1:

$$alb = \sum_{n=1}^{m} (A_n \cdot alb_n) \bigg/ \sum_{n=1}^{m} A_n \quad \text{e.g., albedo.} \tag{2.1}$$

Here, A_n is the area fraction, and alb_n is the albedo for land use category n, while m is the number of categories.

2.2.3 Case Study Setting

Land use data provided by the Geospatial Information Authority of Japan were used to represent the progress of urbanization. Three datasets were used which had been published in 1976, 1987, and 1997, respectively. Since each dataset comprised different land use categories, we recategorized the types of land use into eight new categories and set surface parameters, as shown in Table 2.1.

Figure 2.3a shows the fraction of land in the "urban" category (shaded in Table 2.1) in 1976. The resolution of this figure is 1 km, and the dark tone indicates the fraction of "urban" land use within a 1-km^2 mesh. Figure 2.3b and c shows the incremental increase in the "urban" land use fraction over time. Overall, the results obtained for the two decades targeted did not show extremely rapid urbanization in the Tokyo metropolitan area. Especially in Tokyo's 23 wards, the incremental increase in "urban" land use was very small because Tokyo had already been extensively urbanized by the 1970s. However, the "urban" land use area has increased in the area surrounding Tokyo as urban sprawl has progressed.

In this study, we aimed to evaluate the climatological effect of this urbanization process over the two decades targeted for investigation by using three land use datasets.

Table 2.1 Land use categories and surface parameter settings

Land use categories in the original datasets			Albedo (%)	Moisture availability (%)	Emissivity (% at 9 μm)	Roughness length (cm)	Thermal inertia (cal cm^{-2} K^{-1} s$^{-1/2}$)
Case 76	Case 87	Case 97					
Rice field	Rice field	Rice field	18	50	98.5	10	4
Crop field	Crop field	Crop field	17	30	98.5	15	4
Orchard	Orchard	Forest	13	30	97	50	4
Other tree plantation	Other tree plantation						
Forest	Forest						
Barren land	Barren land	Barren land	25	2	90	1	2
Building site	Building site	Building site	15	10	88	80	3
Road and rail	Road and rail	Road and rail					
Other land use	Other land use	Other land use					
Lake and marsh	Inland water	River basin, lake, and marsh	8	100	98	0.01	6
River basin	Coastal water body	Coastal water body					
Coastal water body							
Beach	Beach	Beach	19	15	92	12	3
–	–	Golf course	19	15	96	12	3

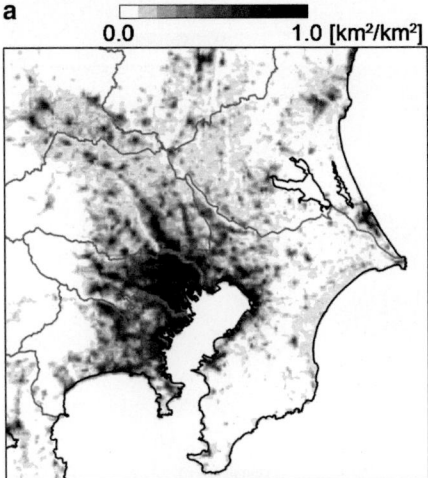

"urban" land use fraction in 1976

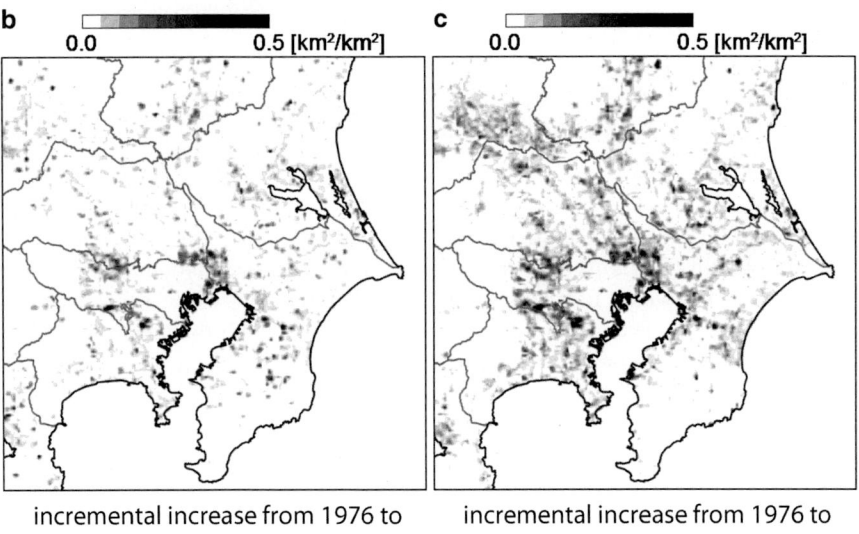

incremental increase from 1976 to   incremental increase from 1976 to
1987   1997

Fig. 2.3 Changes in "urban" land use: (**a**) "urban" land use fraction in 1976 (5,098.099 km² in this domain, 14.272%), (**b**) incremental increase in "urban" land use from 1976 to 1987 (5,567.568 km², 15.586%), and (**c**) incremental increase in "urban" land use from 1976 to 1997 (5,964.726 km², 16.698%)

For simplicity, these three datasets are abbreviated as "case 76" for the land use dataset for 1976, and "case 87" and "case 97" for the 1987 and 1997 datasets, respectively. In order to evaluate the influence of urbanization, the same meteorological datasets were adopted as the initial conditions for all cases investigated.

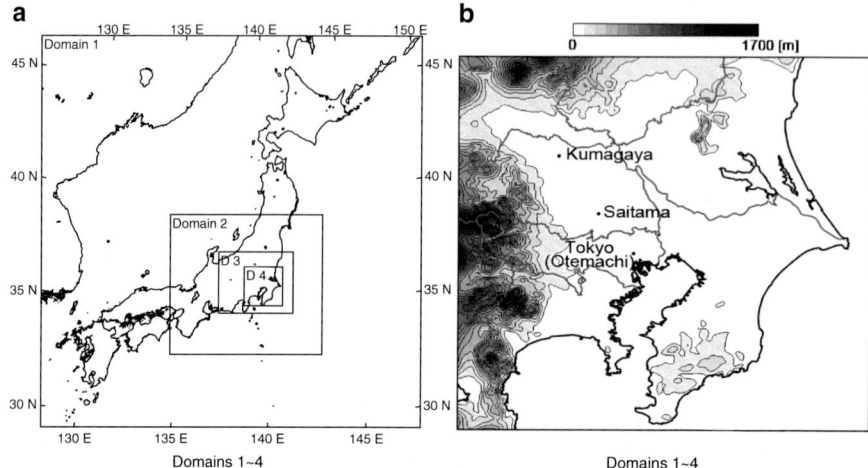

Fig. 2.4 Analysis domains: the four analysis domains are shown in (**a**), while the three observation sites are shown in (**b**), along with the corresponding AMeDAS site from which data were obtained

2.2.4 Analysis Conditions

Figure 2.4 illustrates the analysis domains used, which covered 1,890 km (east–west) × 1,890 km (south–north). The whole analysis domain (domain 1 with a resolution of 27 km) was two-way nested into twofold subdomains: domain 2 (756 × 675 km with a resolution of 9 km) and domain 3 (369 × 297 km with a resolution of 3 km). The land use dataset for 1987 was used for domains 1, 2, and 3. Domain 4 (189 × 189 km with a resolution of 1 km) was one-way nested from the coarse domain and analyzed as three separate cases using three different land use datasets: case 76, case 87, and case 97. The only difference between these three cases was the land use dataset used. The vertical domain was divided into 29 unequally spaced grids. The top was set at approximately 15 km, and the lowest grid height was approximately 36 m.

The thermal environment on July 31, 1986, was then analyzed. On this day, large-scale field observations had been conducted by Yoshikado and Kondo (1989), and their results were compared with the simulation (as shown in later sections). The weather on the target day was categorized as a "frontal type" sea breeze. This means that the temperature profile, wind speed, and wind direction all showed a clear change when the sea breeze front reached inland areas. The synoptic pressure pattern at 09:00 on July 31, 1986, is shown in Fig. 2.5.

The simulations began at 9:00 a.m. (JST) on July 30, and time integration was performed for 39 hours. The initial conditions were given by the JRA-25 dataset (Onogi et al. 2007), and sea surface temperature (SST) was given by NCEP FNL (final) Operational Global Analysis data. Relaxation lateral boundary conditions

Fig. 2.5 Synoptic pressure
pattern on July 31, 1986

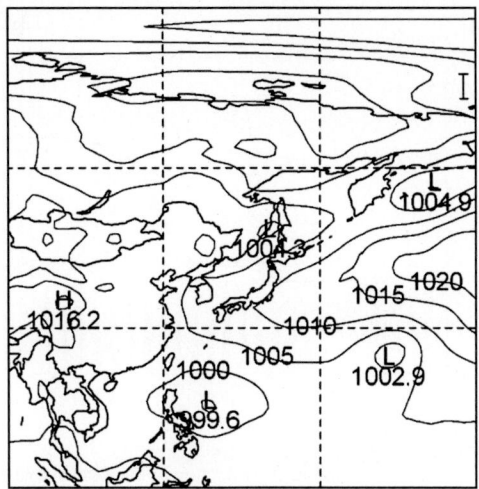

were applied to domain 1, and nest lateral boundary conditions were applied to domains 2, 3, and 4. In order to set the lower boundary conditions, surface temperature was given by the JRA-25 dataset, and SST was given by the NCEP reanalysis dataset. In order to set the upper boundary conditions, radiative conditions were used. In this study, the following schemes were applied for all domains and all cases: (1) Planetary Boundary Layer scheme—Eta PBL scheme (Janjic 1990), (2) Radiation scheme—Cloud-Radiation scheme (Dudhia 1989), and (3) the Surface scheme—Five-Layer Soil model (Dudhia 1996), modified to represent land use fraction.

2.3   Results and Discussion

2.3.1   Results of Base Case (Case 87)

To validate the performance of the mesoscale analysis, the results of case 87 were first compared with the field observations of Yoshikado and Kondo (1989). Figures 2.6 and 2.7 show the time-height cross section of the horizontal wind vector. Figure 2.6 shows the result for the Tokyo observation point, and Fig. 2.7 shows that for the Saitama observation point. The locations of all observation points are shown in Fig. 2.4b. The domain used for this evaluation was domain 4, with a resolution of 1 km.

The simulation result for the Tokyo observation point shows that wind velocity near the surface increased and wind direction shifted to the southeast at 08:00. The dashed line in Fig. 2.6 indicates the change in wind direction along with sea breeze penetration (indicated by "S"). This change in the wind field reflects the penetration of the sea breeze from Tokyo Bay at 08:00. On the other hand,

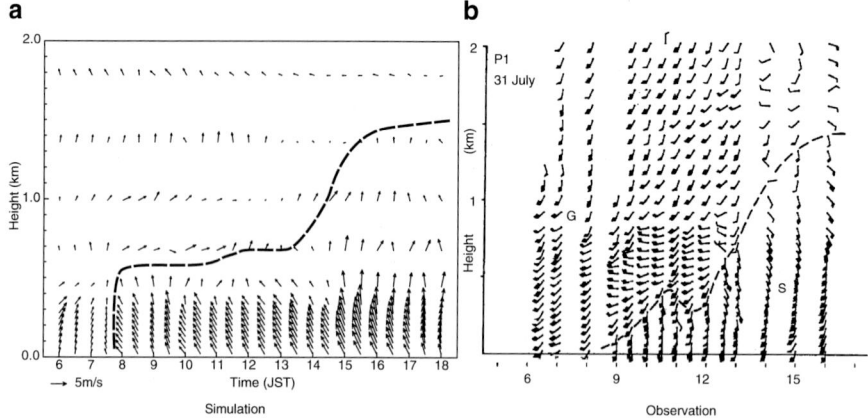

Fig. 2.6 Time-height cross section of the wind above Tokyo: (**a**) simulation result, (**b**) observation result (each wind barb indicates 1 m/s; each pennant indicates 5 m/s) (Yoshikado and Kondo 1989)

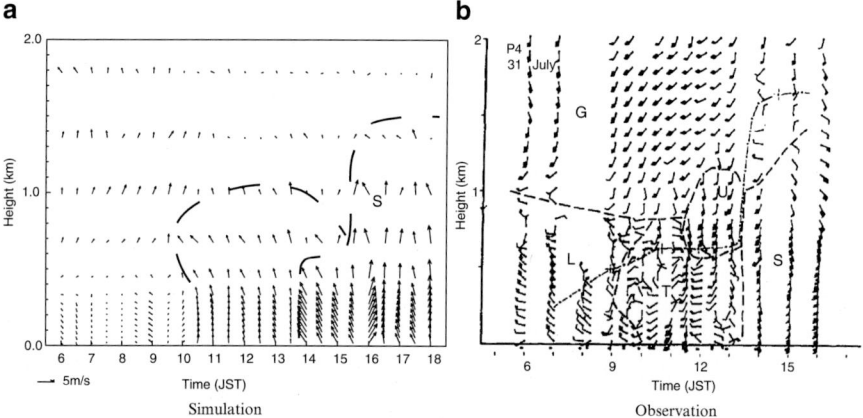

Fig. 2.7 Time-height cross section of the wind above Saitama: (**a**) simulation result, (**b**) observation result (each wind barb indicates 1 m/s; each pennant indicates 5 m/s) (Yoshikado and Kondo 1989)

the observation results showed that this wind field change occurred at 09:00. The driving force affecting the sea breeze is the temperature difference between the air over the sea surface and the land surface. Therefore, it is suspected that the error in sea breeze penetration time was caused by estimation of the SST. As the Tokyo observation point was located near the coastline, it can be assumed that the SST of Tokyo Bay strongly affected the sea breeze penetration. In this study, the SST was set using NCEP FNL data, without taking diurnal variation into consideration. Furthermore, the resolution of the NCEP FNL SST dataset is not fine enough to accurately represent the SST distribution inside Tokyo Bay. These factors could have contributed to the error in estimating the starting time for the penetration of the sea breeze from Tokyo Bay.

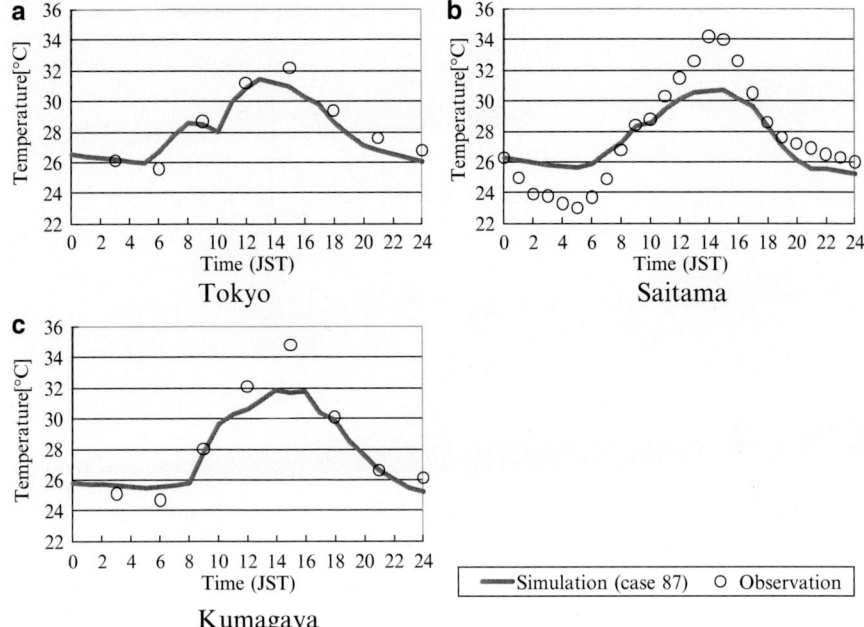

Fig. 2.8 Air temperature time series variation for simulation and observation results at 2 m AGL: (**a**) Tokyo, (**b**) Saitama, (**c**) Kumagaya

Meanwhile, at the Saitama observation site (an inland part of the Tokyo metropolitan area), strong winds appeared at 14:00 in both the simulation and observation results. These results suggest that the sea breeze flowing from Tokyo Bay arrived at that time.

The next step in validating the performance of the mesoscale analysis was to compare the simulation results with AMeDAS (Automated Meteorological Data Acquisition System) observation data collected by the Japan Meteorological Agency. Figure 2.8 shows air temperature time series variation at 2 m above ground level (AGL) for 3 observation points. There were 26 AMeDAS observation stations in domain 4, and three stations were selected for examination. The locations of these observation points are shown in Fig. 2.4b. The domain used for the evaluation was domain 4, with a resolution of 1 km. Both simulation results and observation results were hourly data, averaged every 10 min.

At the Tokyo observation point, the simulation result was a good representation of the actual air temperature variation. However, at the Saitama and Kumagaya observation points, the simulation results underestimated the actual daytime observations. Especially at the Saitama observation point, the range of diurnal air temperature variation in the simulation was very small compared with actual observations. One possible reason for these errors is the representativeness of the results. The target mesh in which the Saitama observation site was located comprises approximately 40% paddy fields, 30% inland waterways, 20% building sites,

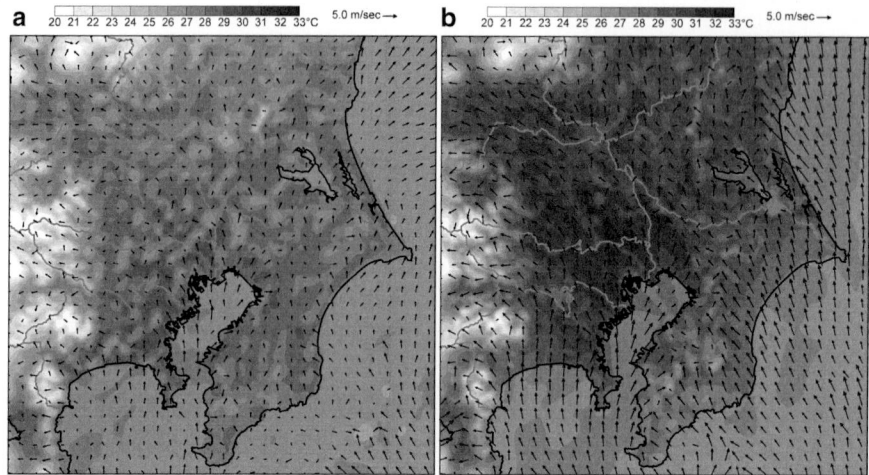

Fig. 2.9 Air temperature and wind velocity vector distribution (air temperature is at 2 m AGL, and the wind velocity vector is at 10 m AGL): (**a**) 09:00 (JST), (**b**) 14:00 (JST)

and small percentages of other land use categories, as shown in case 87. In this study, the surface parameters used in the mesoscale analysis were averaged and weighted by area in order to represent the different land use fractions in each analysis mesh. Because the actual observation site was located at a building site, this averaging process may not have been representative and could be considered the most important factor contributing to the error noted.

Figure 2.9 shows the horizontal distributions of air temperature (2 m AGL) and wind velocity (10 m AGL) in the analysis results at 09:00 and 14:00. The domain used for the evaluation was domain 4, with a resolution of 1 km.

At 09:00, the sea breeze from Tokyo Bay started to flow into coastal areas, while the wind field of inland areas was still weak. In the morning, the range of temperatures across the Kanto Plain was not large (the light-colored area in the west is a mountainous area). By 14:00, the sea breeze from Tokyo Bay had penetrated to the Saitama observation site. By that time, a high-temperature area had also spread out from the central part of Tokyo in a northwest direction and had reached Saitama prefecture. This is a typical pattern for the UHI phenomenon in the Tokyo metropolitan area, and the simulation results represented it well.

2.3.2   Results of Case Study

The results of the variation study used to examine the effects of urbanization are shown below. Note that only land use change was considered in this study. Anthropogenic heat release and urban morphology were not considered because of the lack of detail in the dataset used.

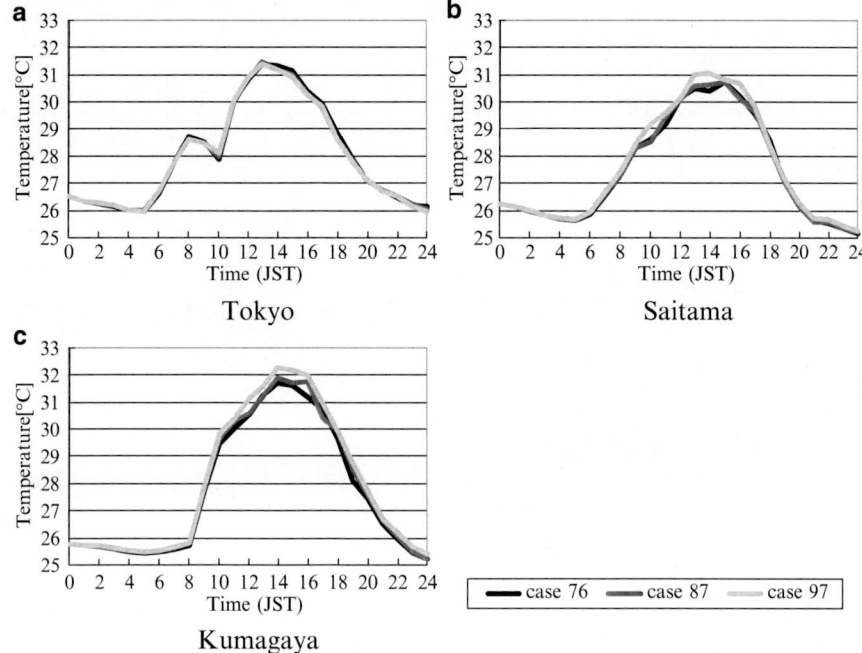

Fig. 2.10 Air temperature time series variation for each case at 2 m AGL: (**a**) Tokyo, (**b**) Saitama, (**c**) Kumagaya

Figure 2.10 shows time series results for air temperature at the three sites used for the three simulation cases. The domain used for the evaluation was domain 4, with a resolution of 1 km. The time interval was 60 min, and all plots were averaged every 10 min.

Although there was only a small difference between each case at each site, a comparatively large temperature rise attributable to urbanization could be seen in inland areas. The maximum difference between cases 76 and 97, around noon, was 0.68°C for Saitama and 0.77°C for Kumagaya. These results confirm that the UHI phenomenon in inland areas has become more serious as a result of the urbanization process. On the other hand, the effect of urbanization on the UHI phenomenon in coastal areas is not large since Tokyo Bay acts as an inlet for cool air.

Figure 2.11 shows wind velocity time series variation. The domain used for the evaluation was domain 4, with a resolution of 1 km. The time interval was 10 min, and plots were averaged every 10 min.

As previously noted, the sea breeze flowing from Tokyo Bay reached the Saitama site at 14:00 in case 87, and this simulation result showed good agreement with actual observations. The results for case 76 showed almost the same trends as case 87, with the dashed line in Fig. 2.9b indicating sea breeze penetration time. However, in case 97, the sea breeze reached the observation site about 20 min later than in cases 76 and 87 (the sea breeze penetration time in case 97 is indicated by the dashed-dotted line in Fig. 2.9b). This result suggests that urbanization in the

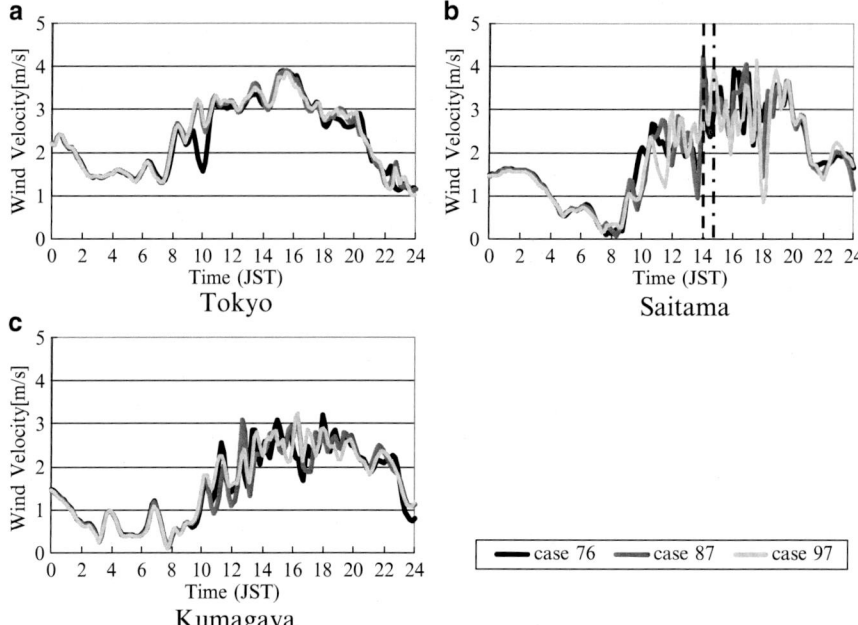

Fig. 2.11 Wind velocity time series variation for each case at 10 m AGL: (**a**) Tokyo, (**b**) Saitama, (**c**) Kumagaya

Tokyo metropolitan area has had an impact not only on the temperature field but also on the wind field. One possible factor preventing the sea breeze penetrating to inland areas is the UHI circulation, with the amplification of UHI intensity strengthening convection over the urban area and perhaps helping to dissipate the momentum of the sea breeze. Another possible reason is the difference in resistance encountered over the urban area. However, the roughness parameter for the "urban" land use category was assumed to be constant in this study. Therefore, the difference in resistance in areas with a high-roughness parameter was not considered a dominant factor.

In this study, the effect of urbanization was examined on the basis of land use change alone. However, there are many other factors that are also changing as part of the urbanization process. These include the increase in anthropogenic heat release to the atmosphere, changes in urban morphology, and so on. Therefore, further investigation into the effects of urbanization on the UHI phenomenon is still needed.

2.4 Conclusion

The progress of the UHI phenomenon in the Tokyo metropolitan area was investigated using the MM5 mesoscale meteorological model. To examine the effects of urbanization, land use datasets for 1976, 1987, and 1997 (provided by the Geospatial Information Authority of Japan) were used. Furthermore, the MM5 system was modified in order to represent land use fractions in each analysis mesh because the original MM5 land surface parameters were based on dominant land use.

The simulated results for case 87, when compared with the observations conducted by Yoshikado and Kondo (1989), predicted an earlier sea breeze penetration time than observations from coastal areas. On the other hand, the arrival time of the sea breeze at inland sites showed good agreement with actual observations.

The results of the variation study carried out in order to examine the effects of urbanization showed small but important differences in temperature and wind velocity diurnal variation in inland areas. The difference between cases 76 and 87 was rather slight, but comparing case 97 with case 76 showed a maximum air temperature difference around noon of 0.68°C and a delay in sea breeze arrival time of about 20 min at the Saitama site.

In this study, the urbanization process was represented by land use change alone. Since urbanization also involves changes in the intensity of anthropogenic heat release, urban morphology, the surface properties of buildings, and so on, the dataset used as input for the mesoscale analysis in this study was not sufficiently detailed. Further investigation into the effects of urbanization on the UHI phenomenon is, therefore, still required.

Appendix

Basic Equations of MM5

In the nonhydrostatic model, pressure p, temperature T, and density of air ρ are defined as follows:

$$p(x, y, z, t) = p_0(z) + p'(x, y, z, t) \tag{2.2}$$

$$T(x, y, z, t) = T_0(z) + T'(x, y, z, t) \tag{2.3}$$

$$\rho(x, y, z, t) = \rho_0(z) + \rho'(x, y, z, t) \tag{2.4}$$

Here, subscript 0 means reference state variables and prime ($'$) means perturbation values.

The terrain-following vertical coordinate is defined by the following equation:

$$\sigma = \frac{p_0 - p_t}{p_{s0} - p_t} \tag{2.5}$$

where p_t is pressure at the top of the simulation domain, p_s is the surface pressure, and p_{s0} is the reference state surface pressure (Fig. 2.12).

The pressure at any arbitrary grid point is given by the following equation:

$$p^*(x, y) = p_s(x, y) - p_t \tag{2.6}$$

$$p = p^*\sigma + p_t + p' \tag{2.7}$$

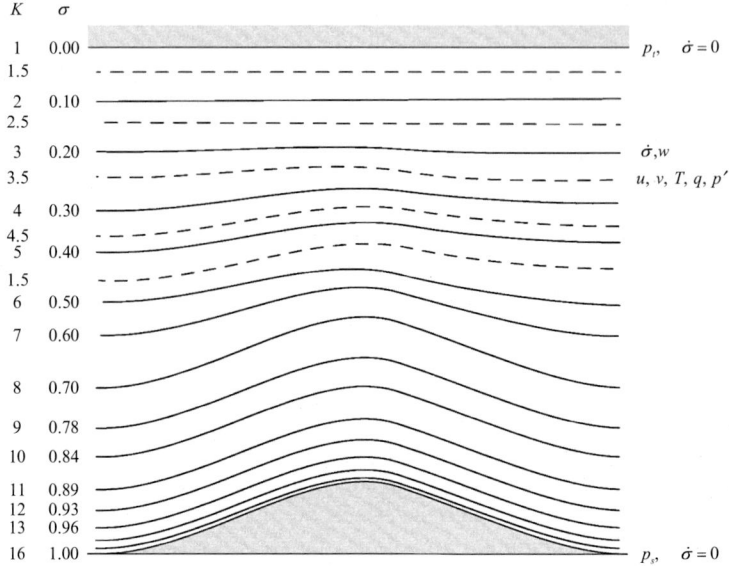

Fig. 2.12 Schematic representation of the vertical structure of the model, showing 15 vertical layers. *Dashed lines* denote half-sigma levels, *solid lines* denote full-sigma levels (Redrawn from Dudhia et al. 2005)

1. Horizontal momentum:

$$\frac{\partial p^* u}{\partial t} = -m^2 \left[\frac{\partial p^* uu/m}{\partial x} + \frac{\partial p^* vu/m}{\partial y} \right] - \frac{\partial p^* u\dot{\sigma}}{\partial \sigma} + uDIV$$
$$- \frac{mp^*}{\rho} \left[\frac{\partial p'}{\partial x} - \frac{\sigma}{p^*} \frac{\partial p^*}{\partial x} \frac{\partial p'}{\partial \sigma} \right] + p^* fv + D_u \qquad (2.8)$$

$$\frac{\partial p^* v}{\partial t} = -m^2 \left[\frac{\partial p^* uv/m}{\partial x} + \frac{\partial p^* vv/m}{\partial y} \right] - \frac{\partial p^* v\dot{\sigma}}{\partial \sigma} + vDIV$$
$$- \frac{mp^*}{\rho} \left[\frac{\partial p'}{\partial y} - \frac{\sigma}{p^*} \frac{\partial p^*}{\partial y} \frac{\partial p'}{\partial \sigma} \right] - p^* fu + D_v \qquad (2.9)$$

2. Vertical momentum:

$$\frac{\partial p^* w}{\partial t} = -m^2 \left[\frac{\partial p^* uw/m}{\partial x} + \frac{\partial p^* vw/m}{\partial y} \right] - \frac{\partial p^* w\dot{\sigma}}{\partial \sigma} + wDIV$$
$$+ p^* g \frac{\rho_0}{\rho} \left[\frac{1}{p^*} \frac{\partial p'}{\partial \sigma} + \frac{T_v'}{T} \frac{T_0 p'}{T p_0} \right] - p^* g[(q_c + q_r)] + D_w \qquad (2.10)$$

3. Pressure:

$$
\begin{aligned}
\frac{\partial p^* p'}{\partial t} = {} & - m^2 \left[\frac{\partial p^* up'/m}{\partial x} + \frac{\partial p^* vp'/m}{\partial y} \right] - \frac{\partial p^* p' \dot{\sigma}}{\partial \sigma} + p' DIV \\
& - m^2 p^* \gamma p \left[\frac{\partial u/m}{\partial x} - \frac{\sigma}{mp^*} \frac{\partial p^*}{\partial x} \frac{\partial u}{\partial \sigma} + \frac{\partial v/m}{\partial y} - \frac{\sigma}{mp^*} \frac{\partial p^*}{\partial y} \frac{\partial v}{\partial \sigma} \right] \\
& + \rho_0 g \gamma p \frac{\partial w}{\partial \sigma} + p^* \rho_0 g w
\end{aligned}
\tag{2.11}
$$

4. Temperature:

$$
\begin{aligned}
\frac{\partial p^* T}{\partial t} = {} & - m^2 \left[\frac{\partial p^* uT/m}{\partial x} + \frac{\partial p^* vT/m}{\partial y} \right] - \frac{\partial p^* T \dot{\sigma}}{\partial \sigma} + T \cdot DIV \\
& + \frac{1}{\rho c_p} \left[p^* \frac{Dp'}{Dt} - \rho_0 g p^* w - D_{p'} \right] + p^* \frac{\dot{Q}}{c_p} + D_T
\end{aligned}
\tag{2.12}
$$

Here, f is the Coriolis parameter, m is the map scale factor, and

$$
DIV = m^2 \left[\frac{\partial p^* u/m}{\partial x} + \frac{\partial p^* v/m}{\partial y} \right] + \frac{\partial p^* \dot{\sigma}}{\partial \sigma}
\tag{2.13}
$$

$$
\dot{\sigma} = - \frac{\rho_0 g}{p^*} w - \frac{m\sigma}{p^*} \frac{\partial p^*}{\partial x} u - \frac{m\sigma}{p^*} \frac{\partial p^*}{\partial y} v
\tag{2.14}
$$

For more details, please refer to Grell et al. (1995).

References

Anthes RA, Warner TT (1978) Development of hydrodynamic models suitable for air pollution and other mesometeorological studies. Mon Weather Rev 106:1045–1078. doi:10.1175/1520-0493(1978) 106<1045:DOHMSF>2.0.CO;2

Dudhia J (1989) Numerical study of convection observed during the winter monsoon experiment using a mesoscale two-dimensional model. J Atmos Sci 46:3077–3107. doi:10.1175/1520-0469(1989) 046<3077:NSOCOD>2.0.CO;2

Dudhia J (1996) A multi-layer soil temperature model for MM5. The Sixth PSU/NCAR mesoscale model users' workshop, pp 49–50

Dudhia J, Gill D, Manning K, Wang W, Bruyere C (2005) PSU/NCAR mesoscale modeling system tutorial class notes and user's guide: MM5 modeling system version 3. Mesoscale and Microscale Meteorology Division, National Center for Atmospheric Research. National Center for Atmospheric Research, Boulder

Fukuoka Y, Nakagawa K (2010) Nairikutoshi ha Naze Atsuika. Seizando, Tokyo (in Japanese). ISBN 978-4-425-51261-4

Grell GA, Dudhia J, Stauffer DR (1995) A description of the fifth-generation Penn State/NCAR mesoscale model (MM5). NCAR TECHNICAL NOTE, TN-398+STR. National Center for Atmospheric Research, Boulder

Gross G (1992) Results of supercomputer simulations of meteorological mesoscale phenomena. Fluid Dyn Res 10:483–489. doi:10.1016/0169-5983(92)90035-U

Howard L (2007) The climate of London. International Association for Urban Climate

Janjic ZI (1990) The step-mountain coordinate: physical package. Monthly Weather Rev 118:1429–1443. doi:10.1175/1520-0493(1990)118<1429:TSMCPP>2.0.CO;2

Kawamoto Y, Ooka R (2008) Improvement of parameterization of ground surface and incorporation of anthropogenic heat release – development of urban climate analysis model using MM5 part 1. J Environ Eng Trans AIJ 631:1125–1132 (in Japanese)

Landsberg HE, Maisel TN (1972) Micrometeorological observation in an area of urban growth. Bound-Lay Meteorol 2:365–370. doi:10.1007/BF02184776

Mochida A, Murakami S, Ojima T, Kim SJ, Ooka R, Sugiyama H (1997) CFD analysis of mesoscale climate in the Greater Tokyo area. J Wind Eng Ind Aerodynam 67:459–477. doi:10.1016/S0167-6105(97)00060-3

Myrup LO (1969) A numerical model of the urban heat island. J Appl Meteorol 8:908–918. doi:10.1175/1520-0450(1969) 008<0908:ANMOTU>2.0.CO;2

Narita K (2006) Ventilation path and urban climate. Wind Eng JAWE 31:109–114 (in Japanese)

Oke TR (1987) Boundary layer climates, 2nd edn. Routledge, London

Onogi K, Tsutsui J, Koide H, Sakamoto M, Kobayashi S, Hatsushika H, Matsumoto T, Yamazaki N, Kamahori H, Takahashi K, Kadokura S, Wada K, Kato K, Oyama R, Ose T, Mannoji N, Taira R (2007) The JRA-25 reanalysis. J Meteorol Soc Jpn 85:369–432. doi:10.2151/jmsj.85.369

Ooka R, Sato T, Murakami S (2008) Numerical simulation of sea breeze over the Kanto plane and analysis of the interruptive factors for the sea breeze based on mean kinetic energy balance. J Environ Eng Trans AIJ 632:1201–1207 (in Japanese)

Orlanski I (1975) A rational subdivision of scales for atmospheric processes. Bull Am Meteorol Soc 56:527–530

Pielke RA, Cotton WR, Walko RL, Tremback CJ, Lyons WA, Grasso LD, Nicholls ME, Moran MD, Wesley DA, Lee TJ, Copeland JH (1992) A comprehensive meteorological modelling system—RAMS. Meteorol Atmos Phys 49:69–91. doi:10.1007/BF01025401

Saito K, Ishida J, Aranami K, Hara T, Segawa T, Narita M, Honda Y (2007) Nonhydrostatic atmospheric models and operational development at JMA. J Meteorol Soc Jpn 85B:271–304. doi:10.2151/jmsj.85B.271

Skamarock WC, Klemp JB, Dudhia J, Gill DO, Barker DM, Duda MG, Huang XY, Wang W, Powers JG (2008) A description of the advanced research WRF Version 3. NCAR Technical Note, NCAR/TN-475 + STR. National Center for Atmospheric Research, Boulder

Sundborg Å (1950) Local climatological studies of the temperature conditions in an urban area. Tellus 2:221–231. doi:10.1111/j.2153-3490.1950.tb00333.x

United Nations, Department of Economic and Social Affairs Population Division, Population Estimates and Projections Section (2009) World urbanization prospects: the 2009 revision. http://esa.un.org/unpd/wup/index.htm. Accessed Jun 17, 2011

World Resources Institute, Fisheries (2007) Population within 100 km of coast. http://earthtrends.wri.org/searchable_db/index.php?theme=1&variable_ID=63&action=select_countries. Accessed Jun 17, 2011

Yoshikado H, Kondo H (1989) Inland penetration of the sea breeze over the suburban area of Tokyo. Bound-Lay Meteorol 48:389–407. doi:10.1007/BF00123061

Chapter 3
Thermal Adaptation Outdoors and the Effect of Wind on Thermal Comfort

Hom Bahadur Rijal

Abstract People use outdoor spaces for various reasons. It is important to create the optimum wind environment to achieve adaptive thermal comfort in a hot and humid climate. To clarify thermal adaptation in the outdoor environment, thermal comfort surveys are reviewed and compared. Semi-outdoor spaces, which have a similar function to the outdoor environment, are also included in the analysis. To clarify the effect of the wind on thermal comfort, indoor thermal comfort surveys are reviewed. The results show that the outdoor comfort temperature is highly correlated with the monthly mean outdoor temperature, indicating the existence of regional and seasonal differences in comfort temperature. The proposed adaptive model can be used to predict the outdoor comfort temperature. Increased wind velocity raises the comfort temperature especially in hot and humid climates.

Keywords Outdoor thermal comfort • Semi-outdoor thermal comfort • Adaptive model • Comfort temperature • Running mean temperature • Wind velocity

3.1 Introduction

People use outdoor spaces for various activities, such as relaxing, meeting people, taking short breaks, and playing. If the outdoor environment is well designed and thermally comfortable, people use it in their daily life. However, we often encounter uncomfortable outdoor environments which need improving by optimizing shading, greenery, open spaces, etc. Especially, if we can improve the wind environment by city planning, people may find they are comfortable both outdoors and indoors.

H.B. Rijal (✉)
Department of Environmental & Information Studies, Tokyo City University,
3-3-1 Ushikubo-nishi, Tsuzuki-ku, Yokohama 224-8551, Japan
e-mail: rijal@tcu.ac.jp

S. Kato and K. Hiyama (eds.), *Ventilating Cities: Air-flow Criteria for Healthy and Comfortable Urban Living*, Springer Geography, DOI 10.1007/978-94-007-2771-7_3, © Springer Science+Business Media B.V. 2012

Thus, ventilating cities and buildings is one of the most important ways to achieve adaptive thermal comfort in a hot and humid climate.

It is true that people need to move from one place to another using outdoor space whether it is comfortable or not. The outdoor environment changes frequently and people are likely to be exposed to different environments during a day, a week, or a month. People amble about, walk quickly, or even run outdoors, which boosts their metabolic rate. People also need to move from comfortable indoor environments to the hot or cold outdoors and vice versa. Because of seasonal and regional adaptation, people may have different thermal expectations of the outdoor environment, which might be an important factor in explaining outdoor thermal comfort. How people adapt to various outdoor conditions is not yet fully understood. If people adapt to a range of outdoor environments, it could also be advantageous to provide some seasonal variation in indoor temperature.

In order to show the variation of the outdoor comfort temperature, a worldwide literature review of field surveys of thermal comfort has been conducted. Semi-outdoor spaces were also included in the review because they have a function similar to that of the outdoor spaces. To predict the outdoor comfort temperature, an adaptive thermal comfort model is proposed. Finally, the effect of wind speed on indoor and outdoor thermal comfort is discussed.

3.2 Outdoor Thermal Comfort

3.2.1 Thermal Comfort Surveys in Europe

Nikolopoulou et al. (2001) conducted an outdoor thermal comfort survey in Cambridge, UK. They investigated at four sites including urban squares, streets, or parks. They used a 5-point thermal sensation scale, varying from −2 (too cold) to +2 (too hot) (0 is neutral). They interviewed 1,431 people in summer and winter. The average comfort temperature was calculated for each interview day. They obtained the following linear regression equation to predict the comfort temperature T_c (°C):

$$T_c = 0.624T_a + 8.0 \ (R^2 = 0.89) \tag{3.1}$$

where T_a is the mean air temperature (°C) for the day's interviews, as measured at the interview site, and R^2 is the coefficient of determination and is based on the daily batches of data.

Nikolopoulou and Lykoudis (2006) report thermal comfort surveys at 14 different case-study sites, across five different European countries: Greece (Athens, Thessaloniki), Italy (Milan), Switzerland (Fribourg), Germany (Kassel), and UK (Cambridge, Sheffield). This was a collaborative project, with a research team from each country, and was called Rediscovering the Urban Realm and Open Spaces (RUROS). The database consists of nearly 10,000 questionnaire-guided interviews. The field surveys were conducted in summer, autumn, winter, and spring. The thermal sensation was collected using a 5-point scale: −2 very cold, −1 cool,

Table 3.1 Regression equation for each season in Europe

Season	Equation	R^2
Summer	$T_c = 1.039T_a + 0.8$	0.82
Autumn	$T_c = 0.568T_a + 12.6$	0.26
Winter	$T_c = 0.550T_a + 13.6$	0.25
Spring	$T_c = 1.024T_a + 6.3$	0.65

0 neither cool nor warm, 1 warm, and 2 very hot. They found the following linear regression equation relating the comfort temperature and climatic air temperature, recorded at the time and place of the interviews, for all the year-round:

$$T_c = 0.507T_a + 12.6 \ (R^2 = 0.49) \tag{3.2}$$

They also calculated an equation for each season, as shown in Table 3.1. If the outdoor temperatures were 10°C in winter and 25°C in summer, the comfort temperatures would be 19.1°C and 26.7°C, respectively. The result shows that the comfort temperature had a large seasonal variation.

Nicol et al. (2006) conducted an outdoor thermal comfort survey in Northern and Southern England (Manchester and Lewes). The thermal sensation vote was collected using the 7-point Bedford scale.[1] The relationship between thermal sensation vote, TSV, and the air temperature at the time and place of the interview is:

$$TSV = 0.136T_a + 1.56 \ (n = 418, \ R^2 = 0.43) \tag{3.3}$$

where n is the number of observations. R^2 is based on the individual observations rather than using grouped data. They obtained the following multiple regression equation to predict the thermal sensation:

$$TSV = 0.132T_a + 0.001I - 0.432\sqrt{v} + 1.8 \ (n = 395, \ R^2 = 0.47) \tag{3.4}$$

where I is solar radiation (W/m^2) and $\sqrt{v}$ is square root of the wind velocity (m/s) 0.5. The square root was used because a number of previous studies had shown that it is a good indication of the cooling effect of air movement (Nicol 1974). The equation suggests that thermal sensation vote will decrease with increasing wind velocity. The comfort temperature was calculated using Griffiths' method (Griffiths 1990), for which they assumed a constant regression coefficient of 0.5 between thermal sensation vote and temperature (Humphreys et al. 2007). The following linear regression equation was obtained to predict the outdoor comfort temperature:

$$T_c = 0.810T_a + 3.6 \ (R^2 = 0.93) \tag{3.5}$$

Equations 3.1, 3.2, and 3.5 are compared in Fig. 3.1. The outdoor comfort temperatures are quite different when outdoor temperature is low but similar

---

[1] The Bedford scale is: 1 much too cool, 2 too cool, 3 comfortably cool, 4 neither warm nor cool, 5 comfortably warm, 6 too warm, and 7 much too warm.

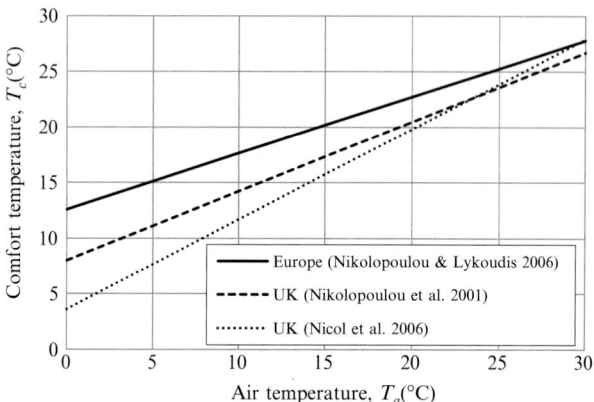

Fig. 3.1 Relation between outdoor comfort temperature and air temperature year-round in UK and Europe

Table 3.2 Questionnaires for thermal comfort survey

Scale	Wind force perception	Wind force preference	Overall comfort or comfort level
5	Very strong	Stronger	Very comfortable
4	Strong	Slightly stronger	Comfortable
3	Acceptable	Acceptable	Acceptable
2	Weak	Slightly weaker	Uncomfortable
1	Very weak	Weaker	Very uncomfortable

when outdoor temperature is high. This difference might be attributable to the use of different thermal sensation scales, to different methods of calculating the comfort temperature, or to cultural differences related to the climatic differences between the regions. Humphreys (1976) had earlier found seasonal differences in indoor comfort temperature that were related to the climate. The results show that this is also true of outdoor comfort temperatures.

Finally, Metje et al. (2008) conducted an outdoor thermal comfort survey in Birmingham, UK. A total of 451 sets of responses were obtained in summer and winter. The questionnaires used in the field survey are shown in Table 3.2. To evaluate the overall comfort (OC) (from air temperature, wind velocity, relative humidity, and solar radiation), a 5-point scale was used. The following regression equation is found to predict the overall comfort:

$$OC = 0.066T_a + 2.5 \ (R^2 = 0.88) \tag{3.6}$$

$$OC = -0.253v + 3.5 \ (R^2 = 0.87) \tag{3.7}$$

The equations are shown in Fig. 3.2. The comfort value increases with increasing air temperature and reduces with increasing wind velocity. Because of the cool climate in UK, the strong wind may have negative effect for thermal comfort.

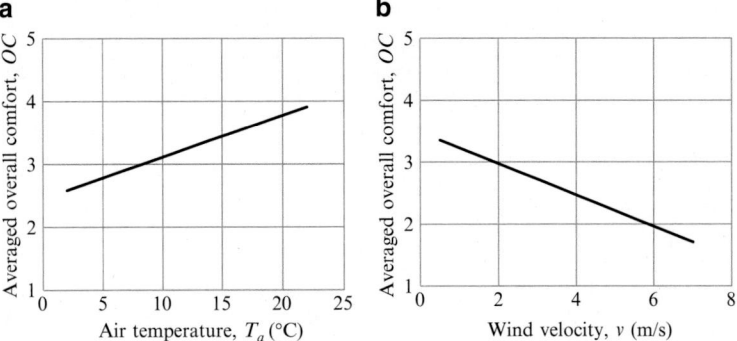

Fig. 3.2 Relation between overall comfort and (**a**) air temperature and (**b**) wind velocity in Birmingham, UK

The number of people who rated the wind velocity as uncomfortable and very uncomfortable was more than 60%, for wind velocity greater than 5 m/s. Penwarden (1973) had earlier found a similar value for the tolerable wind velocity. Tacken (1989) suggested that the average wind velocity should not be higher than the 2.5 m/s for designing areas suited to outdoor relaxation in the Dutch climate. At lower wind velocity, approximately 80% of the respondents rated the comfort level as acceptable or comfortable.

In Metje's survey, wind speed above about 2 m/s was regarded as "strong" or "very strong." More than 60% of the respondents felt that the wind was strong or very strong for wind velocity greater than approximately 3.5 m/s.

Metje et al. (2008) obtained the following multiple regression equation to predict the overall comfort:

$$OC = 0.067T_a - 0.226v - 0.008H_r + 0.001I + 3.2 \ (R^2 = 0.25) \quad (3.8)$$

where v is wind velocity (m/s), H_r is relative humidity (%), and I is solar radiation (W/m^2). By considering the most important factors influencing the overall comfort, they simplified in the following equation:

$$OC = 0.076T_a - 0.182v + 2.6 \ (R^2 = 0.23) \quad (3.9)$$

In this research, the overall comfort was used as a dependent variable. Overall comfort would be expected to have a curvilinear relation to the air temperature—one would expect there to be an optimum temperature for comfort, so the use of linear regression gives an incomplete description of the relation. The meaning of the overall comfort scale would change according to season, and so, it is difficult to interpret the results. For example, "very uncomfortable" can be "heat discomfort" in summer or "cold discomfort" in winter. To find the optimum comfort condition, curvilinear multiple regression would need to be applied. Most other researchers used the thermal sensation, rather than overall comfort, as a dependent variable. Thus, this research is not directly comparable with other researches.

Table 3.3 The questionnaires for the outdoor comfort survey

Questions	5-point scale				
(a) Perception: considering the time and season, the air temperature is high	−2 Disagree	−1	0 Uncertain	1	2 Agree
(b) Preference: considering your activity and clothing, it would be (more) comfortable, should the air temperature be	−2 Lower	−1	0 Unchanged	1	2 Higher

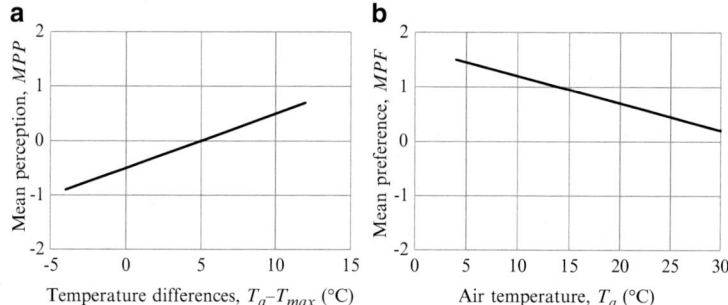

Fig. 3.3 Relation between (**a**) perception and T_a–T_{max} and (**b**) preference and T_a in Montreal, Canada

3.2.2 Thermal Comfort Survey in Canada

Stathopoulos et al. (2004) conducted an outdoor thermal comfort survey in Montreal, Canada. The data were collected in seven public spaces in the downtown area mostly in spring and autumn. A total of 466 responses were obtained by interview using a 5-point scale as shown in Table 3.3. They obtained the following regression equation using daily average data to predict the mean perception, *MPP*, and the mean preference, *MPF*:

$$MPP = 0.1(T_a - T_{max}) - 0.5 \ (r = 0.58) \tag{3.10}$$

$$MPF = -0.05T_a + 1.7 \ (r = -0.78) \tag{3.11}$$

where T_{max} is the daily mean maximum temperature (°C) for the month, derived from the monthly norms of long-term meteorological records. The regression lines are shown in Fig. 3.3. They found that very few people preferred a higher wind velocity, a higher relative humidity, or a lower solar radiation, regardless of the actual weather conditions. The perception was related to the temperature difference ($T_a - T_{max}$). Because of the generally cool to cold climate in Montreal, the mean preference is greater than zero.

Fig. 3.4 Relation between
the *TSV* and *SET** *in* Fukuoka
and Tokyo, Japan

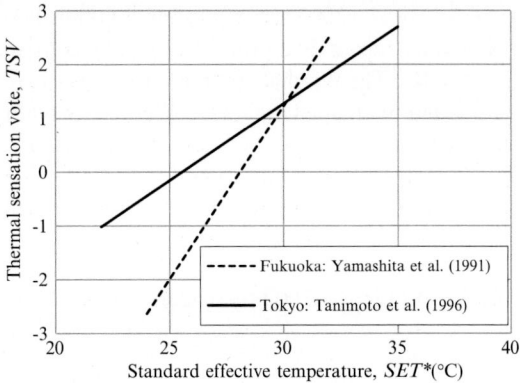

3.2.3   Thermal Comfort Survey in Japan

Tanimoto et al. (1996) conducted an outdoor thermal comfort survey in summer in Tokyo, Japan. They constructed two rest-facility buildings for experiments. Both buildings have four pillars and roof, but no walls. Sixteen university students participated in the experiments. The *TSV* were collected using 7-point rating scale indicating −3 cold, −2, −1, 0 neither hot nor cold, 1, 2, 3 hot. The air temperature, relative humidity, globe temperature, and wind velocity were measured. The following regression equation was found in between the *TSV* and Standard Effective Temperature,[2] *SET** (°C):

$$TSV = 0.286SET^* - 7.31 \qquad (3.12)$$

Tanimoto et al. (1996) referred to the work of Yamashita et al. (1991) who had found the following regression equation for outdoor thermal comfort in a survey in Fukuoka (southern Japan):

$$TSV = 0.642SET^* - 18.04 \qquad (3.13)$$

The *TSV* was collected using 7-point ASHRAE scale.[3] Equations 3.12 and 3.13 are shown in Fig. 3.4. When, *TSV* = 0, the comfort SET* is 25.6°C in Tokyo and 28.1°C in Fukuoka. The comfort *SET** in Fukuoka is 2.5°C higher than in Tokyo. The reason might be that the monthly mean outdoor temperature of Fukuoka was 26.7°C which is 1.9°C higher than Tokyo, and thus people may have adapted to the higher ambient temperatures in Fukuoka.

---

[2] The *SET** is calculated using air temperature, relative humidity, mean radiant temperature, wind velocity, metabolic rate, and clothing insulation. It represents the temperature that would have felt the same had the clothing been 0.5 clo and the activity 1 met rather than the actual values.

[3] The ASHRAE scale is: −3 cold, −2 cool, −1 slightly cool, 0 neutral, 1 slightly warm, 2 warm, and 3 hot.

Givoni et al. (2003) conducted an outdoor thermal comfort survey in summer in Yokohama, Japan. The group of subjects consisted of six people including males and females. The thermal comfort survey was conducted in an open area exposed to the sun and in an open area with a vertical wind break. They used a 7-point thermal sensation scale indicating 1 (very cold) to 7 (very hot). The air temperature, ground surface temperature, relative humidity, wind velocity, and horizontal solar radiation were measured. The following multiple regression equation was obtained to predict the thermal sensation vote:

$$TSV = 0.112T_a + 0.002I - 0.322v - 0.007H_r + 0.005T_s + 1.7 \ (R^2 = 0.88) \quad (3.14)$$

where H_r is relative humidity (%) and T_s is surrounding ground surface temperature (°C). They simplified regression equation by considering the relative effects of air temperature, solar radiation, and wind velocity on thermal comfort and removing less influential variables:

$$TSV = 0.112T_a + 0.002I - 0.319v + 1.2 \ (R^2 = 0.87) \quad (3.15)$$

As they mentioned, the equation is based on a very small number of people, and thus it should be used only for the indication of the regression model. Further investigation is required for a fully acceptable conclusion to be reached.

3.2.4 Thermal Comfort Survey in Taiwan

In order to account for tourists' thermal perception, Lin and Matzarakis (2008) conducted a year-round outdoor thermal comfort survey in Taiwan. Thermal sensation votes (TSV) were collected from 1,644 tourists by interview using the 7-point ASHRAE scale. The physiologically equivalent temperature,[4] PET (°C), was used for the evaluation. The mean thermal sensation vote (MTSV) was calculated for each 1°C PET interval. They found the following regression equation in between MTSV and PET:

$$MTSV = 0.0559PET - 1.52 \ (R^2 = 0.83, \ p<0.001) \quad (3.16)$$

---

[4] The physiological equivalent temperature (PET) is based on the Munich Energy-balance Model for Individuals (MEMI), which models the thermal conditions of the human body in a physiologically relevant way (Höppe 1999). PET is defined as the air temperature at which, in a typical indoor setting (without wind and solar radiation), the heat budget of the human body is balanced with the same core and skin temperature as under the complex outdoor conditions to be assessed. This way, PET enables a layperson to compare the integral effects of complex thermal conditions outside with his or her own experience indoors. Lin and Matzarakis (2008) calculated the PET using air temperature, globe temperature, relative humidity, wind velocity, metabolic rate, and clothing insulation.

Fig. 3.5 Relation between
the *MTSV* and *PET* in Taiwan

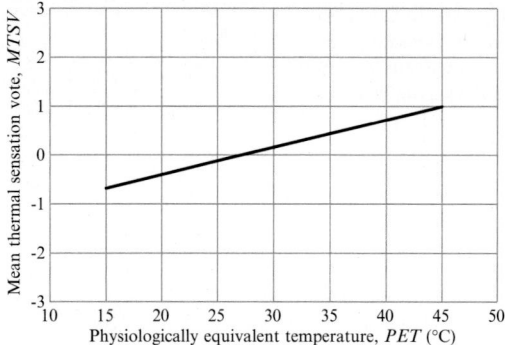

where *p* is level of significance. The equation is illustrated in Fig. 3.5. When *MTSV* is neutral (0), the comfort temperature is 27.2°C *PET*. The value of *PET* calculated to be neutral in Western/Middle Europe is 18–23°C *PET* (Matzarakis and Mayer 1996). The difference indicates that people tolerate a higher temperature in Taiwan's tropical climate.

3.2.5   Thermal Comfort Survey in Israel

Givoni et al. (2003) conducted an outdoor thermal comfort survey in an urban park in Tel Aviv, Israel. The range of the air temperature during the two measurements days was 23–27°C. The subjects-group consisted of 10 persons including males and females. Thermal sensation votes were collected in one shaded area and five sun-exposed areas using a scale of 0 very cold to 9 unbearable (4 is neutral). They found the following multiple regression equation to predict the mean thermal sensation vote:

$$MTSV = 0.22T_a - 0.05v + 0.003I - 2.3 \qquad (3.17)$$

Givoni et al. (2003) also conducted an outdoor thermal comfort survey in a Kibbutz in Israel. The thermal measurements were taken during four consecutive days in areas in the sun and in the shade. The maximum daily air temperature varied from 34°C to 41°C in 4 days. Fourteen people participated in the thermal comfort survey. The thermal sensation scale is the same which is used in the Tel Aviv. They obtained the following linear regression equation for sun and shade areas:

Sun $$\qquad TSV = 0.1697T_a + 1.27 \; (R^2 = 0.47) \qquad (3.18)$$

Shade $$\qquad TSV = 0.1882T_a - 0.584 \; (R^2 = 0.57) \qquad (3.19)$$

Sun $$\qquad TSV = 0.003I + 5.1 \; (R^2 = 0.71) \qquad (3.20)$$

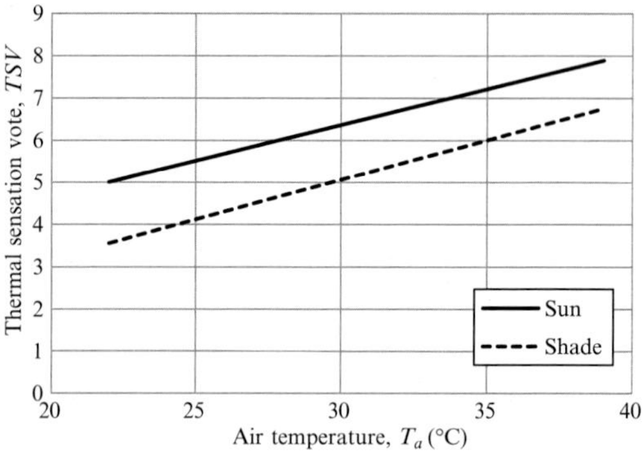

Fig. 3.6 Relation between thermal sensation vote and air temperature for subjects in the sun or shaded in Kibbutz, Israel

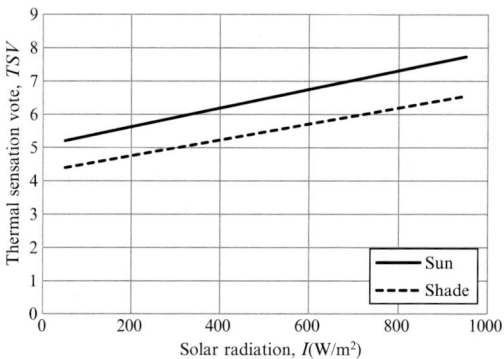

Fig. 3.7 Relation between thermal sensation vote and solar radiation for subjects in the sun or shaded in Kibbutz, Israel

Shade $\qquad\qquad TSV = 0.002I + 4.3\ (R^2 = 0.50)$ $\qquad\qquad$ (3.21)

These equations are shown in Figs. 3.6 and 3.7. As expected, thermal sensation in the shade is lower (cooler) than in the sun. When thermal sensation vote is 5, the comfort temperature is 22.0°C in sun and 29.7°C in shade. The comfort temperature of the shaded area is 7.7°C higher than the sunny area. The results showed that the shade area is important to create the comfortable thermal environment in this climate.

3.3 Semi-outdoor Thermal Comfort

3.3.1 Thermal Comfort Survey in Japan

Chun and Tamura (1998) conducted a thermal comfort survey in underground shopping malls and department stores in Yokohama, Japan. The underground shopping malls are more influenced by outdoor climate than are the department

Fig. 3.8 *Regression line* for semi-outdoor spaces year-round in Tokyo, Japan

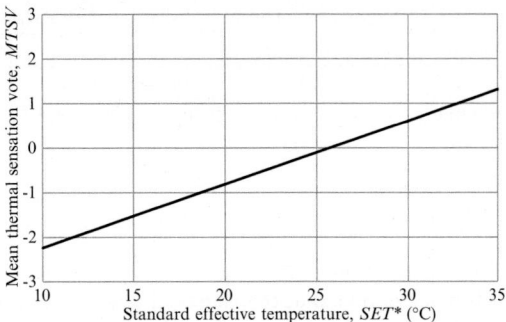

stores, especially in the passageways. Japanese underground shopping malls are neither completely closed spaces nor completely open spaces; they are semi-open spaces. For this chapter, only the results from the underground shopping malls are relevant. Thermal sensation votes were collected from 16 university students using an 11-point scale: 1 very cold, 2 cold, 3 slightly cold, 4 cool, 5 slightly cool, 6 neutral, 7 slightly warm, 8 warm, 9 slightly hot, 10 hot, and 11 very hot. The following regression equations were obtained from *TSV* and air temperature:

August $$TSV = 0.45T_a - 5.6 \ (r = 0.70) \tag{3.22}$$

October $$TSV = 0.47T_a - 5.9 \ (r = 0.80) \tag{3.23}$$

December $$TSV = 0.24T_a + 0.5 \ (r = 0.80) \tag{3.24}$$

The comfort temperatures are 25.6°C in August, 24.8°C in October, and 21.5°C in December. The result showed that the comfort temperature in summer was higher than in winter.

Nakano and Tanabe (2004) conducted a thermal comfort survey in four semi-outdoor spaces in Tokyo, Japan. They collected the 2,248 sets of thermal sensation votes, using the 7-point ASHRAE scale and measured thermal environmental data, in naturally ventilated and air-conditioned atria in four seasons. For this chapter, only the two naturally ventilated semi-outdoor spaces were selected. To calculate the *SET** (standard effective temperature), measured clothing insulation was used. However, the metabolic rate was assumed to be 1.1 met for all respondents. Calculated values of *SET** were rounded into 1.0°C increments, and corresponding *MTSV* were derived. The following weighted regression equation was found for year-round:

$$MTSV = 0.142SET^* - 3.65 \ (R^2 = 0.92) \tag{3.25}$$

The equation is shown in Fig. 3.8. When *MTSV* = 0 (neutral), the comfort *SET** is 25.7°C. A seasonal difference was found in the comfort *SET** (Table 3.4).

Table 3.4 Regression equations for semi-outdoor spaces for each season

Season	Equation	R^2	SET^*_c (°C)
Summer	$MTSV = 0.1845SET^* - 4.9615$	0.88	26.9
Autumn	$MTSV = 0.2078SET^* - 4.8663$	0.80	23.4
Winter	$MTSV = 0.1357SET^* - 3.3848$	0.65	24.9
Spring	$MTSV = 0.1741SET^* - 4.1687$	0.83	23.9

SET^*_c comfort SET^*

SET^* compensates for differences in clothing insulation, so the remaining seasonal difference shows that the seasonal adaptation was greater than could be explained by clothing changes alone. This suggests that physiological or psychological adaptation had occurred.

3.3.2 Thermal Comfort Survey in Thailand

Jitkhajornwanich et al. (1998) conducted a thermal comfort survey in transitional spaces during the cool season in Bangkok, Thailand. Indoor and outdoor transitional spaces were investigated including the entrance halls, lobby areas, foyers, or under canopies of the buildings. The outdoor transitional spaces included shaded areas such as under trees and under canopies. Thus this research is included in the semi-outdoor spaces. Thermal sensation votes were collected from 593 subjects using 7-point ASHRAE scale. The following regression equation was found in between $MTSV$ and air temperature:

$$MTSV = 0.21T_a - 5.6 \ (R^2 = 0.75) \tag{3.26}$$

The comfort temperature in the transitional spaces was 27.1°C which is similar to that obtained for the naturally ventilated buildings in Bangkok.

3.3.3 Thermal Comfort Survey in Nepal

Rijal et al. (2010) conducted a thermal comfort survey in semi-outdoor spaces of traditional houses, during summer and winter, in three districts of Nepal. They collected 3,000 thermal comfort votes using 9-point thermal sensation scale: −4 very cold, −3 cold, −2 cool, −1 slightly cool, 0 neutral, 1 slightly warm, 2 warm, 3 hot, and 4 very hot. The globe temperature was measured using 150-mm diameter globe. The comfort temperature was calculated using Griffiths' method (Griffiths 1990; Rijal et al. 2008), taking the regression coefficient as 0.33. The following equation was found between the comfort temperature and mean globe temperature at the time of voting, T_{gm}:

$$T_c = 0.765T_{gm} + 6.1 \ (R^2 = 0.998) \tag{3.27}$$

Fig. 3.9 Thermal comfort survey in semi-outdoor spaces in temperate and subtropical climates of Nepal

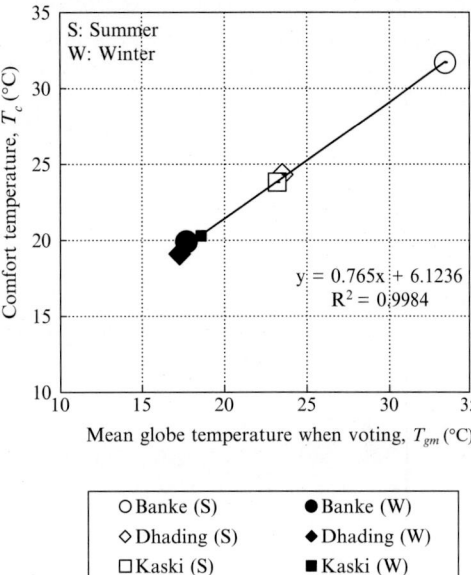

The results are shown in Fig. 3.9. The comfort temperature is highly correlated with the mean globe temperature and has large seasonal differences. In summer, the comfort temperature in the subtropical climate (Banke district) is significantly higher than in a temperate climate (Dhading and Kaski districts). If the globe temperature is high, the comfort temperature tends to be high, which suggested that the higher ambient temperature raises the comfort temperature.

3.4   An Adaptive Comfort Model for Outdoors

This section brings together the results of the outdoor and semi-outdoor thermal comfort studies that have been described above. To formulate an adaptive model for outdoor thermal comfort, the relation between the comfort temperature and monthly mean outdoor temperature is formulated. The outdoor comfort temperatures are from Europe, Israel, Japan, Thailand, Nepal, and Australia, which gives a fair coverage of worldwide climates. The comfort temperature and outdoor temperature are shown in Table 3.5. The comfort temperatures were extracted from the papers or calculated using the regression equation. Comfort temperatures are expressed in terms of the air temperature or the globe temperature or SET^* or PET. Long-term monthly mean outdoor temperatures, T_o (°C), were obtained from the nearest meteorological station or those reported in the paper were used. From these data (Table 3.5), a regression analysis was conducted for outdoor and semi-outdoor spaces as shown in Fig. 3.10. The equation of the regression lines are:

$$\text{Outdoor } T_c = 0.620T_o + 10.7 \ \left(n = 39, \ R^2 = 0.67, \ SE = 0.07, \ p{<}0.001\right) \quad (3.28)$$

Table 3.5 Comfort temperature and outdoor temperature

Spaces	References	Country	City	Period	Meteorological station	Variable	T_o (°C)	T_c (°C)
Outdoor	Nikolopoulou and Lykoudis (2006)	Greece	Athens	Summer	Nearest station	T_a	27.0	28.5
	Nikolopoulou and Lykoudis (2006)	Greece	Thessaloniki	Summer	Nearest station	T_a	25.7	28.9
	Nikolopoulou and Lykoudis (2006)	Switzerland	Fribourg	Summer	Nearest station	T_a	16.8	15.8
	Nikolopoulou and Lykoudis (2006)	Italy	Milan	Summer	Nearest station	T_a	22.0	21.5
	Nikolopoulou and Lykoudis (2006)	UK	Cambridge	Summer	Nearest station	T_a	16.3	18.0
	Nikolopoulou and Lykoudis (2006)	UK	Sheffield	Summer	Nearest station	T_a	15.7	15.8
	Nikolopoulou and Lykoudis (2006)	Germany	Kassel	Summer	Nearest station	T_a	16.6	22.1
	Nikolopoulou and Lykoudis (2006)	Greece	Athens	Autumn	Nearest station	T_a	19.7	19.4
	Nikolopoulou and Lykoudis (2006)	Greece	Thessaloniki	Autumn	Nearest station	T_a	16.3	24.7
	Nikolopoulou and Lykoudis (2006)	Switzerland	Fribourg	Autumn	Nearest station	T_a	8.8	13.2
	Nikolopoulou and Lykoudis (2006)	Italy	Milan	Autumn	Nearest station	T_a	13.0	24.6
	Nikolopoulou and Lykoudis (2006)	UK	Cambridge	Autumn	Nearest station	T_a	10.7	23.2
	Nikolopoulou and Lykoudis (2006)	UK	Sheffield	Autumn	Nearest station	T_a	10.2	16.7
	Nikolopoulou and Lykoudis (2006)	Germany	Kassel	Autumn	Nearest station	T_a	9.0	15.8
	Nikolopoulou and Lykoudis (2006)	Greece	Athens	Winter	Nearest station	T_a	11.0	21.5
	Nikolopoulou and Lykoudis (2006)	Greece	Thessaloniki	Winter	Nearest station	T_a	6.3	15.0
	Nikolopoulou and Lykoudis (2006)[a]	Switzerland	Fribourg	Winter	Nearest station	T_a	−0.2	11.9
	Nikolopoulou and Lykoudis (2006)	Italy	Milan	Winter	Nearest station	T_a	2.6	21.1
	Nikolopoulou and Lykoudis (2006)	UK	Sheffield	Winter	Nearest station	T_a	4.4	10.8
	Nikolopoulou and Lykoudis (2006)	Germany	Kassel	Winter	Nearest station	T_a	0.6	15.2
	Nikolopoulou and Lykoudis (2006)	Greece	Athens	Spring	Nearest station	T_a	16.3	24.3
	Nikolopoulou and Lykoudis (2006)	Greece	Thessaloniki	Spring	Nearest station	T_a	14.5	18.4
	Nikolopoulou and Lykoudis (2006)	Switzerland	Fribourg	Spring	Nearest station	T_a	8.6	13.2
	Nikolopoulou and Lykoudis (2006)	Italy	Milan	Spring	Nearest station	T_a	12.4	20.7
	Nikolopoulou and Lykoudis (2006)	UK	Cambridge	Spring	Nearest station	T_a	8.8	17.6
	Nikolopoulou and Lykoudis (2006)	UK	Sheffield	Spring	Nearest station	T_a	8.6	11.8
	Nikolopoulou and Lykoudis (2006)	Germany	Kassel	Spring	Nearest station	T_a	8.1	17.2
	Nicol et al. (2006)	UK	Manchester	Summer	Manchester	T_a	15.3	16.2

	Reference	Country	Location	City	Season		T_o	T_c
	Nicol et al. (2006)	UK	Manchester	Manchester	Winter	T_a	4.8	10.7
	Nicol et al. (2006)	UK	Lewes	Southampton	Summer	T_a	16.3	22.9
	Nicol et al. (2006)	UK	Lewes	Southampton	Winter	T_a	6.0	8.6
	Tanimoto et al. (1996)	Japan	Tokyo	Tokyo	Summer	SET^*	24.8	25.6
	Yamashita et al. (1991)	Japan	Fukuoka	Fukuoka	Summer	SET^*	26.7	28.1
	Nakano and Tanabe (2004)	Japan	Tokyo	Tokyo	Summer	SET^*	26.0	26.9
	Nakano and Tanabe (2004)	Japan	Tokyo	Tokyo	Autumn	SET^*	18.3	23.4
	Nakano and Tanabe (2004)[a]	Japan	Tokyo	Tokyo	Winter	SET^*	7.9	24.9
	Nakano and Tanabe (2004)	Japan	Tokyo	Tokyo	Spring	SET^*	15.6	23.9
	Lin and Matzarakis (2008)	Taiwan	Sun Moon Lake	Sun Moon Lake	One year	PET	19.2	27.2
	Givoni et al. (2003)	Israel	Kibbutz (Sunny area)	Beersheva	July, August	T_a	27.0	22.0
	Givoni et al. (2003)	Israel	Kibbutz (Shaded area)	Beersheva	July, August	T_a	27.0	29.7
	Spagnolo and de Dear (2003)[a]	Australia	Sydney	Sydney	Winter	T_a	11.7	26.6
	Spagnolo and de Dear (2003)	Australia	Sydney	Sydney	Summer	T_a	21.7	23.0
Semi-outdoor	Chun and Tamura (1998)	Japan	Yokohama	Yokohama	August	T_a	26.4	25.6
	Chun and Tamura (1998)	Japan	Yokohama	Yokohama	October	T_a	17.2	24.8
	Chun and Tamura (1998)	Japan	Yokohama	Yokohama	December	T_a	7.7	21.5
	Jitkhajornwanich et al. (1998)	Thailand	Bangkok	Bangkok	Cool season	T_a	26.0	27.1
	Rijal et al. (2010)	Nepal	Banke	Nepaljung	Summer	T_g	31.4	31.7
	Rijal et al. (2010)	Nepal	Banke	Nepaljung	Winter	T_g	15.2	19.9
	Rijal et al. (2010)	Nepal	Dhading	Dhunibesi	Summer	T_g	25.4	24.3
	Rijal et al. (2010)	Nepal	Dhading	Dhunibesi	Winter	T_g	13.3	19.1
	Rijal et al. (2010)	Nepal	Kaski	Lumle	Summer	T_g	18.8	23.8
	Rijal et al. (2010)	Nepal	Kaski	Lumle	Winter	T_g	8.9	20.3

T_o long-term climatic average outdoor temperature (Some data are monthly average of the particular year) (°C), T_c comfort temperature (°C), T_g globe temperature (°C), SET^* Standard Effective Temperature (°C), PET Physiological Equivalent Temperature (°C)

[a]Excluded in the regression analysis

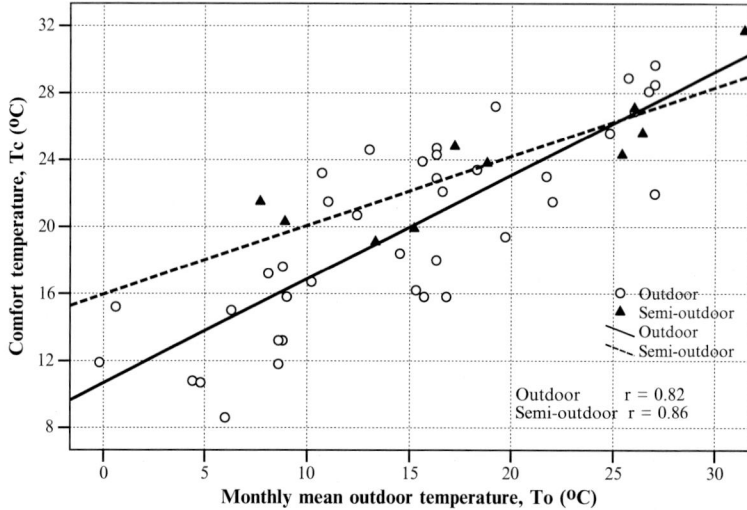

Fig. 3.10 Relation between the comfort temperature and monthly mean outdoor temperature in outdoors or semi-outdoors

Semi - outdoor $T_c = 0.413T_o + 16.0 \left(n = 10, \ R^2 = 0.74, \ SE = 0.09, \ p = 0.001 \right)$

$$(3.29)$$

where SE is the standard error of the regression coefficient.

Nikolopoulou and Lykoudis (2006) also showed this relationship using their European data including the winter comfort temperature for Athens, Milan, and Kassel. They found the following equation for the year-round:

Outdoor $\qquad\qquad T_c = 0.507T_o + 12.6 \ (R^2 = 0.49)$ $\qquad\qquad$ (3.30)

The regression coefficient and correlation coefficient of this study are slightly higher than Nikolopoulou and Lykoudis. The regression coefficient for the outdoor data is significantly higher than for the semi-outdoors ($p = 0.01$). The outdoor comfort temperature is related to the monthly mean outdoor temperature for both semi-outdoor and outdoor environment. Humphreys (1978) found that indoor comfort temperature was related to the monthly mean outdoor temperature, particularly in buildings operating in the free-running mode (no energy being used for heating or cooling). The results showed that Humphreys finding applies in principle to the outdoor or semi-outdoor comfort temperature.

It is sometimes uncertain, for spaces that are very close to buildings, whether to classify them as outdoors or semi-outdoors. It may therefore be appropriate to combine the two kinds of data despite the significant difference between them noted above. To predict the overall outdoor comfort temperature, the data of outdoors and

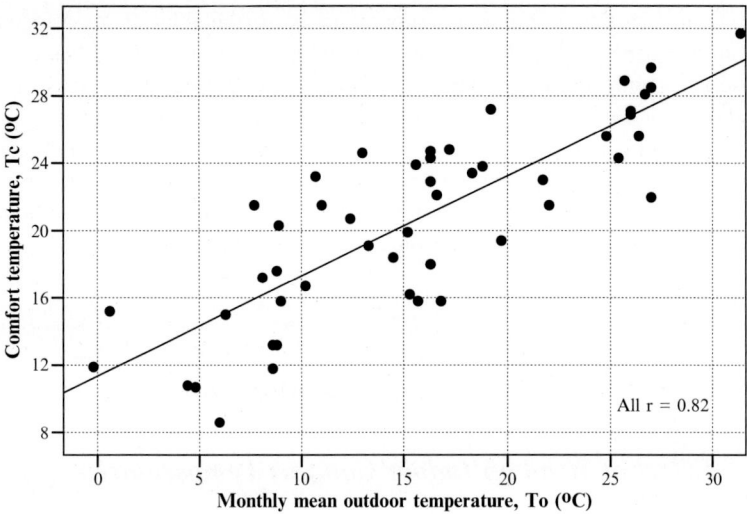

Fig. 3.11 Relation between the comfort temperature and monthly mean outdoor temperature both in outdoors and semi-outdoors

semi-outdoors were combined as shown in the Fig. 3.11. Sixty-eight percent of the variation of the comfort temperature is associated with the variation of the outdoor temperature ($R^2 = 0.68$). The equation of the regression line is:

All $T_c = 0.594T_o + 11.4$ $(n = 49,\ R^2 = 0.68,\ SE = 0.06,\ p{<}0.001)$ (3.31)

To predict the indoor comfort temperate (T_{ci}) in the free-running mode, Humphreys (1978) and Humphreys et al. (2010) found the following regression equations for buildings that were neither being heated nor cooled (the free-running mode):

$$T_{ci} = 0.534T_o + 11.9\ (n = 27,\ R^2 = 0.94)$$ (3.32)

$$T_{ci} = 0.550T_o + 13.2\ (R^2 = 0.64)$$ (3.33)

It is interesting to note that the regression coefficient for outdoors (Eq. 3.31) is similar to those for indoors (Eqs. 3.32 and 3.33). At an outdoor temperature of 25°C, the comfort temperature would be 26.3°C, 25.3°C, and 27.0°C, respectively. The results showed that the indoor and outdoor comfort temperatures are very similar. The reason could be that indoor and outdoor temperature is highly correlated in the free-running mode, and thus comfort temperatures are similar in outdoors and indoors. It seems to imply that people outdoors are adapted to the same temperature as they have indoors in free-running conditions. The equation of all data can be used to predict the most probable outdoor comfort temperature from a knowledge of the monthly mean air temperature at that location. The considerable scatter at any particular outdoor temperature is attributable, at least in part, to different air speeds and intensities of solar radiation among the studies on which the observed comfort temperatures rest. The effect of air speed is discussed in the next section.

The equation is based on rather few surveys, and thus we need to validate it by conducting more outdoor thermal comfort survey in various climates for fully acceptable conclusion. It would be desirable to establish a standard methodology so that comparisons between the various climates could be better made.

3.5 Effects of Wind on Thermal Comfort

3.5.1 Effect of Wind on Indoor Comfort Temperature

In tropical climates, air movement is an important factor in determining the indoor comfort temperature (Nicol 2004). A simple theoretical analysis suggests that where the wind velocity is above 0.1 m/s and fairly constant, an allowance can be made in raising the indoor comfort temperature by (Nicol 2004; Humphreys 1970; Humphreys and Nicol 1995)

$$7 - \frac{50}{4 + 10v^{0.5}}\,^{\circ}C \tag{3.34}$$

The relationship is shown in Fig. 3.12. The result shows that if wind velocity is 1 m/s, the indoor comfort temperature can be increased by some 3°C or 4°C. It means that if the indoor comfort temperature is 28°C without air movement, it can be increased to 31.4°C with 1 m/s wind velocity. However, this result was calculated, assuming a metabolic rate of 1 met, 30% of this being lost by evaporation, and no wind penetration of the clothing. These conditions may not be typical of people in outdoor and semi-outdoor spaces, particularly in hot conditions. So, it is wise to compare the result with empirical data from warm climates.

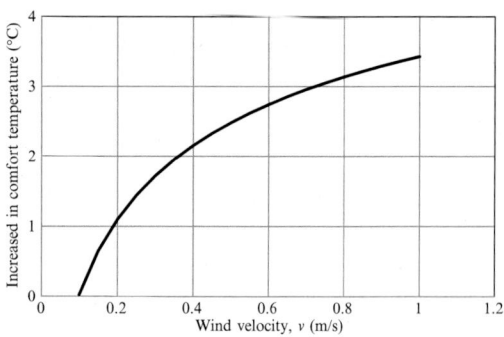

Fig. 3.12 Increase in indoor comfort temperature for different wind velocities

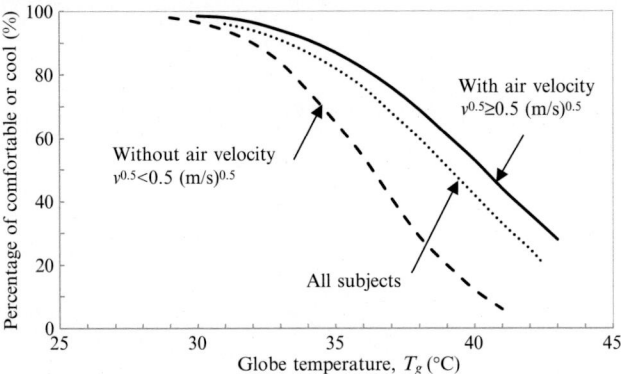

Fig. 3.13 Probabilities of being comfortable or cool against indoor globe temperature in the hot and dry climates of Northern India and Iraq (Nicol 1974)

To clarify the effect of the air movement, Nicol (1974) analyzed indoor thermal comfort data in the hot and dry climates of Northern India and Iraq. More than 2,000 samples were collected from 16 people in homes and offices. Thermal sensation vote was collected using the 7-point Bedford scale. He found that the presence of air movement can be equivalent to a reduction in indoor temperature of as much as 4°C (Nicol 2004) (Fig. 3.13). The major effect on comfort was observed where the square root of wind velocity exceeded 0.5 $(m/s)^{0.5}$ (i.e., wind velocities greater than 0.25 m/s). The effect of wind velocity on reported skin moisture is shown in Fig. 3.14. When indoor air temperature is 31–40°C, the increased wind velocity reduced the skin moisture.

Sharma and Ali (1986) conducted an indoor thermal comfort survey in hot-dry and warm-humid conditions in Roorkee, India. Thermal sensation vote was collected using Bedford scale from 18 young male adults over a period of three consecutive summer seasons. The total number of samples was 5,155. They proposed the following approximate equation to calculate the tropical summer index,[5] *TSI*:

$$TSI = 0.33T_w + 0.75T_g - 2\sqrt{v} \qquad (3.35)$$

where T_w is wet-bulb temperature (°C) and T_g is indoor globe temperature (°C). The *TSI* is highly related with the thermal sensation ($r = 0.82$). They found the effect of wind that the *TSI* decreases when wind velocity is increased (Fig. 3.15).

Nicol et al. (1999) conducted indoor thermal comfort surveys in five climatic zones of Pakistan. The number of samples collected from 33 offices and commercial buildings was over 7,000. Monthly surveys were conducted year-round. Thermal sensation votes were collected using the 7-point Bedford scale. Rijal et al. (2008)

---

[5] The tropical summer index is defined as the air/globe temperature of still air at 50% relative humidity which produces the same overall thermal sensation as the environment under investigation.

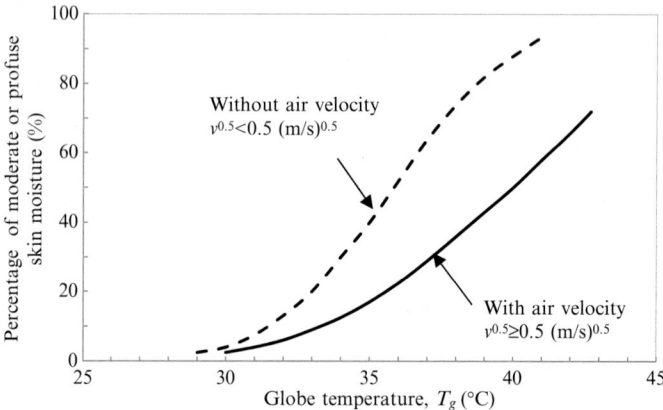

Fig. 3.14 Probabilities of moderate or profuse skin moisture against indoor globe temperature in the hot and dry climates of Northern India and Iraq (Nicol 1974). The scale used for skin moisture is 1 none, 2 slight, 3 moderate, and 4 profuse

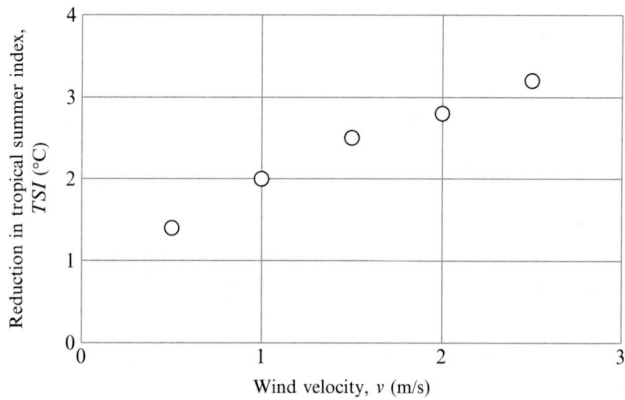

Fig. 3.15 Reduction in *TSI* for different wind velocity in Roorkee, India

analyzed this database and found the following regression equation for ceiling fan use:

Fan off $T_c = 0.408T_{rm} + 16.6$ $(n = 2,769, \; r = 0.69, \; SE = 0.08)$ (3.36)

Fan on $T_c = 0.480T_{rm} + 16.6$ $(n = 2,810, \; r = 0.71, \; SE = 0.09)$ (3.37)

where T_{rm} is the running mean temperature, and the following expression is used for calculation:

$$_nT_{rm} = \alpha_{n-1}T_{rm} + (1 - \alpha) \cdot {}_{n-1}T_{dm}$$ (3.38)

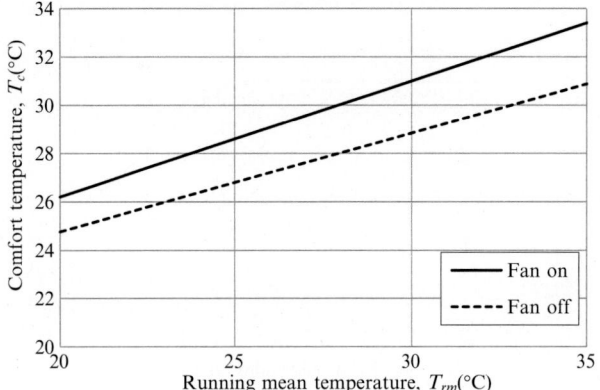

Fig. 3.16 Effect of ceiling fan on indoor comfort temperature in Pakistan

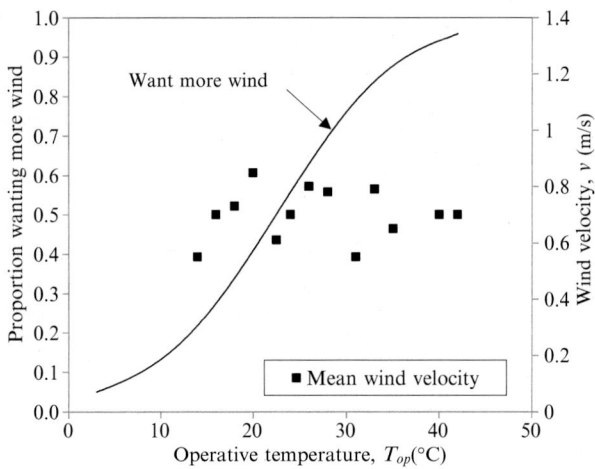

Fig. 3.17 Relation between the operative temperature, wind velocity, and wind velocity preference in Sydney, Australia

where $_nT_{rm}$ is the running mean temperature for day n (°C), $_{n-1}T_{rm}$ is the running mean temperature for the previous day (°C), $_{n-1}T_{dm}$ is daily mean outdoor temperature for the previous day (°C), and α is constant. So, if the running mean temperature has been calculated (or assumed) for one day, then it can be readily calculated for the next day, and so on. A value of 0.80 for α was taken (McCartney and Nicol 2002). This value suggests that the characteristic time subjects take to adjust fully to a change in the outdoor temperature is about a week (Nicol 2008).

Equations 3.37 and 3.38 are shown in Fig. 3.16. Nicol (2004) also showed a similar figure using this database. When running mean temperature is 30°C,

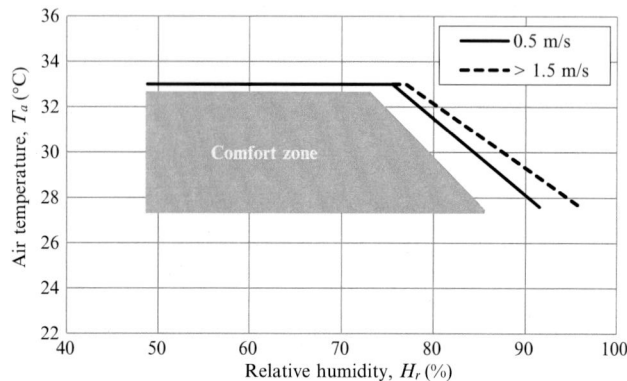

Fig. 3.18 Summer outdoor comfort zone in Dhaka, Bangladesh

the indoor comfort temperature when the fan is on is 31°C which is 2.2°C higher than when the fan is off. When the running mean temperature is high, the fan is perhaps slightly more effective in increasing the indoor comfort temperature, but the difference between the regression coefficients of the two lines is not statistically significant.

Summary: The results of the various studies all confirm that the wind velocity is effective in raising the indoor comfort temperature, and there is quite good agreement about the magnitude of the effect. Generally, wind velocity outdoors is higher than indoors, and thus it might raise still more the outdoor comfort temperature, which would be beneficial in hot and humid climates. However, outdoor thermal comfort is complicated, and thus further research is required to quantify the effect of wind velocity on outdoor thermal comfort.

3.5.2 Wind Preference Outdoors

Spagnolo and de Dear (2003) conducted an outdoor and semi-open-space thermal comfort survey in summer and winter in subtropical Sydney, Australia. They investigated in unshaded and shaded areas including a railway station, a bus interchange, an urban canyon, and a ferry terminal. From the subjects, 1,018 responses were collected using the 7-point ASHRAE scale. The wind preference was asked about using a 3-point scale: more air movement, no change, or less air movement.

The proportion of people wanting more wind increases when operative temperature increases (Fig. 3.17). Increased air velocity is effective for increasing evaporative heat loss at higher temperature and moisture level if the skin is damp. It can be concluded that the people preferred higher air movement outdoors to feel comfortable in a subtropical climate.

3.5.3 Effect of Wind and Humidity on Comfort

Ahmed (2003) conducted an outdoor thermal comfort survey in Dhaka, Bangladesh. Thermal sensation votes were collected in summer from approximately 1,500 people using the 7-point ASHRAE scale. The summer comfort zone is shown in Fig. 3.18. The zone was derived for people involved in activity of 1 met and wearing 0.35–0.50 clo under shaded condition. The shaded area is the comfort zone when there is no wind. As shown in figure, the higher wind velocity raises the upper limit of the comfort zone when the relative humidity is high.

3.6 Concluding Comments

In this chapter, outdoor and semi-outdoor thermal comfort surveys were reviewed, and an adaptive relation proposed to predict the outdoor comfort temperature from the monthly mean outdoor air temperature. To clarify the effect of the wind velocity on thermal comfort, some indoor thermal comfort surveys were reviewed. The following observations are made:

1. The comfort temperatures in surveys of outdoor or semi-outdoor spaces were usually estimated by applying univariate linear regression, or multiple linear regression analysis. The method is adequate provided the variance of the outdoor temperature is substantial and the people interviewed are not far from their comfort temperature.
2. Regional and seasonal differences were found in the outdoor comfort temperature.
3. Outdoor comfort temperatures are highly related to the monthly mean outdoor temperature. The results showed that outdoor comfort temperature is similar to the indoor comfort temperature particularly in buildings operating in the free-running mode (no energy being used for heating or cooling). The proposed adaptive equation can be used to predict the outdoor comfort temperature. The uncertainty in the prediction is probably caused by differences in incident solar radiation and air speed.
4. Wind velocity is effective to increase the indoor comfort temperature especially in the hot and humid climate. Generally, wind velocity outdoors is higher than indoors, and thus it might raise still more the outdoor comfort temperature, which would be beneficial in hot and humid climates.

Nomenclature

H_r	Relative humidity (%)
I	Horizontal solar radiation (W/m^2)
$MTSV$	Mean Thermal Sensation Vote ($-$)
MPP	Mean Perception ($-$)
MPF	Mean Preference ($-$)
OC	Overall Comfort ($-$)
PET	Physiologically Equivalent Temperature ($^\circ$C)
SET^*	Standard Effective Temperature ($^\circ$C)
SET^*_c	Comfort SET^* ($^\circ$C)
T_a	Air temperature ($^\circ$C)
T_o	Monthly mean temperatures ($^\circ$C)
T_g	Globe temperature ($^\circ$C)
T_{gm}	Mean globe temperature when voting ($^\circ$C)
T_w	Wet-bulb temperature ($^\circ$C)
T_c	Comfort temperature ($^\circ$C)
T_{ci}	Indoor comfort temperature ($^\circ$C)
T_{max}	Daily maximum temperature, derived from the monthly norm of long-term meteorological records ($^\circ$C)
T_{rm}	Running mean temperature ($^\circ$C)
$_nT_{rm}$	Running mean temperature for day n ($^\circ$C)
$_{n-1}T_{rm}$	Running mean temperature for the previous day ($^\circ$C)
$_{n-1}T_{dm}$	Daily mean outdoor temperature for the previous day ($^\circ$C)
T_s	Surrounding ground surface temperature ($^\circ$C)
TSI	Tropical Summer Index ($^\circ$C)
TSV	Thermal Sensation Vote ($-$)
v	Wind velocity (m/s)
$\sqrt{v}$	Square root of the wind velocity (m/s)$^{0.5}$
α	Constant ($=0.80$)
n	Number of observations
p	Level of significant
r	Correlation coefficient
R^2	Coefficient of determination

Acknowledgments This chapter is based on a literature review, and I would like thank all researchers whose work I have used. I would like to give especial thanks to Prof. Michael Humphreys and Prof. Fergus Nicol of Oxford Brookes University for their support and encouragement.

References

Ahmed KS (2003) Comfort in urban spaces: defining the boundaries of outdoor thermal comfort for the tropical urban environments. Energ Build 35:103–110

Chun CY, Tamura A (1998) Thermal environment and human responses in underground shopping malls vs department stores in Japan. Build Environ 33(2–3):151–158

Givoni B, Noguchi M, Saaroni H, Pochter O, Yaacov Y, Feller N, Becker S (2003) Outdoor comfort research issues. Energ Build 35:77–86

Griffiths ID (1990) Thermal comfort in buildings with passive solar features: field studies. Report to the Commission of the European Communities, EN3S-090, UK

Höppe P (1999) The physiological equivalent temperature – a universal index for the biometeorological assessment of the thermal environment. Int J Biometeorol 43:71–75

Humphreys MA (1970) A simple theoretical derivation of thermal comfort conditions. J Inst Heating Ventilating Eng 38:95–98

Humphreys MA (1976) Field studies of thermal comfort compared and applied. J Inst Heat Ventilation Eng 44:5–23

Humphreys MA (1978) Outdoor temperatures and comfort indoors. Building Research and Practice. J CIB 6(2):92–105

Humphreys MA, Nicol JF (1995) An adaptive guideline for UK office temperatures. In: Nicol F, Humphreys M, Sykes O, Roaf S (eds) Standards for thermal comfort. E FN Spon, Chapman & Hall, London, pp 190–195

Humphreys MA, Nicol JF, Raja IA (2007) Field studies of indoor thermal comfort and the progress of the adaptive approach. Adv Build Energ Res 1:55–88

Humphreys MA, Rijal HB, Nicol JF (2010) Examining and developing the adaptive relation between climate and thermal comfort indoors. In: Proceedings of conference: Adapting to change: new thinking on comfort, Windsor, UK, Network for Comfort and Energy Use in Buildings, London, 9–11 April 2010

Jitkhajornwanich K, Pitts AC, Malama A, Sharples S (1998) Thermal comfort in transitional spaces in the cool season of Bangkok. ASHRAE Trans Part 1B:1181–1193

Lin TP, Matzarakis A (2008) Tourism climate and thermal comfort in Sun Moon Lake, Taiwan. Int J Biometeorol 52:281–290

Matzarakis A, Mayer H (1996) Another kind of environmental stress: thermal stress. WHO News 18:7–10

McCartney KJ, Nicol JF (2002) Developing an adaptive control algorithm for Europe. Energ Build 34(6):623–635

Metje N, Sterling M, Baker CJ (2008) Pedestrian comfort using clothing values and body temperatures. J Wind Eng Ind Aerodyn 96:412–435

Nakano J, Tanabe S (2004) Thermal comfort and adaptation in semi-outdoor environments. ASHRAE Trans 110(2):543–553

Nicol JF (1974) An analysis of some observations of thermal comfort in Roorkee, India and Baghdad, Iraq. Ann Human Biol 1(4):411–426

Nicol F (2004) Adaptive thermal comfort standards in the hot–humid tropics. Energ Build 36 (7):628–637

Nicol F (2008) Adaptive standards for thermal comfort in buildings. 21–30. In: Adaptive thermal comfort in buildings, The Kinki Chapter of the Society of Heating, Air-conditioning and Sanitary Engineers of Japan, Kyoto, Japan

Nicol JF, Raja IA, Allaudin A, Jamy GN (1999) Climatic variations in comfortable temperatures: the Pakistan projects. Energ Build 30(3):261–279

Nicol F, Wilson E, Ueberjahn-Tritta A, Nanayakkara L, Kessler M (2006) Comfort in outdoor spaces in Manchester and Lewes, UK. In: Proceeding of international conference on comfort and energy use in buildings: getting them right (Windsor), Organised by the Network for Comfort and Energy Use in Buildings

Nikolopoulou M, Lykoudis S (2006) Thermal comfort in outdoor urban spaces: analysis across different European countries. Build Environ 41:1455–1470

Nikolopoulou M, Baker N, Steemers K (2001) Thermal comfort in outdoor urban spaces: understanding the human parameter. Solar Energ 70(3):227–235

Penwarden AD (1973) Acceptable wind speeds in towns. Build Sci 8(3):259–267

Rijal HB, Tuohy P, Humphreys MA, Nicol JF, Samuel A, Raja IA, Clarke J (2008) Development of adaptive algorithms for the operation of windows, fans and doors to predict thermal comfort and energy use in Pakistani buildings. ASHRAE Trans 114(2):555–573

Rijal HB, Yoshida H, Umemiya N (2010) Seasonal and regional differences in neutral temperatures in Nepalese traditional vernacular houses. Build Environ 45(12):2743–2753

Sharma MR, Ali S (1986) Tropical summer index—a study of thermal comfort of Indian subjects. Build Environ 21(1):11–24

Spagnolo J, de Dear R (2003) A field study of thermal comfort in outdoor and semi-outdoor environments in subtropical Sydney Australia. Build Environ 38:721–738

Stathopoulos T, Wu H, Zacharias J (2004) Outdoor human comfort in an urban climate. Build Environ 39:297–305

Tacken M (1989) A comfortable wind climate for outdoor relaxation in urban areas. Build Environ 24(4):321–324

Tanimoto J, Kimura K, Sato T (1996) Experimental study on outdoor rest facilities using passive cooling techniques. J Archit Plann Environ Eng AIJ 481:41–49 (in Japanese with English abstract)

Yamashita M, Ishii A, Iwamoto S, Katayama T, Shiotsuki Y (1991) An experimental study on thermal sensations in the outdoor environment (Part 6) Comparison of temperature sensation with thermal indices. Summaries of Technical Papers of Annual Meeting AIJ (D): 751–752 (in Japanese)

Chapter 4
Health Risk of Exposure to Vehicular Emissions in Wind-Stagnant Street Canyons

Tomomi Hoshiko, Fumiyuki Nakajima, Tassanee Prueksasit, and Kazuo Yamamoto

Abstract In recent years, most stationary sources of air pollution have been removed from urban areas; however, mobile sources are currently the direct causes of air pollution–related health problems. In particular, in the complex configurations of city buildings, the wind becomes stagnant in street canyons, which leads to higher levels of air pollution inside these areas. In this chapter, the health risks due to exposure to vehicular emissions in street canyons are discussed using a case study of field measurement and risk assessment in the street canyons of Bangkok, Thailand. The pollutants of focus are polycyclic aromatic hydrocarbons (PAHs), one of the major hazardous air pollutants from vehicular emissions. Environmental standards have not been introduced in most of the countries, and information on PAH pollution is still lacking in Asian developing countries, where the population densities and levels of traffic pollution are reported to be very high, particularly in the large cities. This chapter also includes a literature review on PAHs. The Bangkok case study of field measurement and risk assessment was conducted for roadside residents who live in possible hot spots of traffic air pollution. The field measurements provide detailed information on PAH levels, such as the diurnal variations and seasonal variations in concentrations, taking influential factors of traffic and wind conditions into consideration. The results of

T. Hoshiko (✉)
Graduate School of Engineering, The University of Tokyo, 7-3-1 Hongo,
Bunkyo-ku, Tokyo 1138656, Japan
e-mail: hoshiko@env.t.u-tokyo.ac.jp

F. Nakajima • K. Yamamoto
Environmental Science Center, The University of Tokyo, 7-3-1 Hongo,
Bunkyo-ku, Tokyo 1110033, Japan
e-mail: nakajima@esc.u-tokyo.ac.jp; yamamoto@esc.u-tokyo.ac.jp

T. Prueksasit
Faculty of Science, Chulalongkorn University, Phayathai Road, Pathumwan,
Bangkok 10330, Thailand
e-mail: tassanee.c@chula.ac.th

S. Kato and K. Hiyama (eds.), *Ventilating Cities: Air-flow Criteria for Healthy and Comfortable Urban Living*, Springer Geography, DOI 10.1007/978-94-007-2771-7_4, © Springer Science+Business Media B.V. 2012

the risk assessment suggested that Bangkok roadside residents in the street canyons are exposed to significant levels of health risk due to PAH exposure.

Keywords Risk assessment • Polycyclic aromatic hydrocarbons • Field measurement • Diurnal and seasonal variations • Bangkok

4.1 Introduction

One important aspect of urban air quality control is its risk assessment for urban dwellers. In recent years, most of stationary sources have been removed from urban areas; however, mobile sources are currently the direct causes of health risk to people. Health risks of exposure to vehicular emissions have long been concerning, especially for those in the vicinity of roads with heavy traffic in cities (e.g., WHO 2002; Pope and Dockery 1999). In street canyons, due to complex building configurations and micrometeorological effects such as stagnant wind, the atmospheric behavior of traffic emissions is complicated, and the health risks due to exposure to air pollutants have not been sufficiently studied. This study focuses on one of the major hazardous air pollutants from vehicular emissions, polycyclic aromatic hydrocarbons (PAHs), for which environmental standards have not been introduced in most countries. Information about PAH pollution is relatively abundant for western countries and Japan; however, it is still lacking in developing Asian countries, where the population densities and levels of traffic pollution are very high in the large cities. This chapter introduces the risk assessment of urban air through a case study in Bangkok, Thailand. Field measurements allow for exposure levels and health risks to be assessed for roadside residents who are living in possible hot spots of urban air pollution. Section 4.2 summarizes the relevant information through a literature review. Section 4.3 presents a case study of field measurements and risk assessment. Conclusions are discussed in Sect. 4.4.

4.2 Atmospheric Behavior and Inhalation Risk of PAHs

Polycyclic aromatic hydrocarbons are emitted through various combustion processes into the atmosphere. The environmental concern for the airborne PAHs comes from the recognition that several of them are potent carcinogens. Currently, European working groups (European Commission 2001) take the lead in stressing the necessity for studying the health effects of atmospheric PAHs and proposing the establishment of risk assessment–based regulations. Risk assessment requires monitoring the concentration levels of important PAHs and identifying their sources. Significant limitations currently exist in the large body of environmental monitoring data for PAHs, including the standardized selection of priority PAHs to be monitored, the analysis methods, and toxicity and risk assessments; however, a store of knowledge on PAHs has only recently been acquired. This chapter reviews previous studies on the sources and atmospheric behavior of PAHs as well as the inhalation risks from these compounds.

4.2.1   Sources and Atmospheric Behavior of PAHs

Polycyclic aromatic hydrocarbons are formed during the incomplete combustion of oil, coal, gas, wood, and other organic substances. PAHs are initially generated in the gas phase, and they are adsorbed on preexisting particles undergoing condensation during further cooling of the emission. Thus, most ambient PAHs exist in the particulate phase, while some higher-volatility PAHs or low molecular weight PAHs remain partly in the gas phase (e.g., Beak et al. 1991).

More than 100 different PAHs have been identified in atmospheric particulate matter and in emissions from coal-fired residential furnaces, and approximately 200 PAHs have been found in environmental tobacco smoke (IPCS 1998). There are basically five major emission source components: domestic, mobile, industrial, agricultural, and natural. The relative importance of these sources changes depending on the place or regulatory views; however, in urban areas, mobile sources are the major contributors (e.g., Nielsen 1996). In Bangkok, Panther et al. (1996) reported that about 88% of PAH emission is attributed to motor vehicles, and minor contributions are from biomass burning and oil combustion. Generally, in the urban areas of several developed countries, PAH concentrations are reported to be in a long-term decreasing trend, which correlates with increased use of catalytic converters in motor vehicles (Beak et al. 1991; European Environment Agency 2004). By contrast, in the large cities of some developing countries, higher levels of PAHs have been observed due to the recent advance of motorization, although monitoring data are still limited (Panther et al. 1999).

The general properties of PAHs include their high boiling and melting points and low vapor pressure. However, the physical and chemical properties of individual PAHs may vary. Table 4.1 summarizes the physicochemical properties of 13 PAHs that were analyzed in this study, and Fig. 4.1 shows their structural formulas. Among the 13 PAHs, 12 of them, not including benzo(e)pyrene (BeP), have been included in the priority pollutant list of the Clean Water Act of the United States Environmental Protection Agency (US EPA) since the 1970s.

The World Health Organization (WHO) has examined the health risks of PAHs on a number of occasions and published Air Quality Guidelines in 1987 and 2000, which recommended the annual average BaP concentration of 1.0 ng/m^3 (WHO 1987, 2000). PAHs are covered by the Persistent Organic Pollutant (POP) Protocol under the United Nations Economic Commission for Europe's Convention on Long-Range Transboundary Air Pollution (UN ECE CLRTAP). Under the protocol, emissions of four PAH compounds must be reported annually, and in addition, the PAH emissions in 2010 may not exceed the levels of the reference year 1990. Some member states have adopted or are considering adopting air quality standards for the selected PAHs. For example, Italy has legally enforceable ambient air

Table 4.1 Physicochemical properties of 13 PAHs (IPCS 1998)

Compound	Abbr.	Molecular formula	Molecular weight	No. of aromatic rings	Vapor pressure (Pa at 25°C)	Solubility in water (μg/L at 25°C)
Phenanthrene	Phe	$C_{14}H_{10}$	178	3	1.6×10^{-2}	1.3×10^3
Anthracene	Ant	$C_{14}H_{10}$	178	3	8.0×10^{-4}	73
Fluoranthene	Fluo	$C_{16}H_{10}$	202	4	1.2×10^{-3}	260
Pyrene	Pyr	$C_{16}H_{10}$	202	4	6.0×10^{-4}	135
Benzo(a)anthracene	BaA	$C_{18}H_{12}$	228	4	2.8×10^{-5}	5.6
Chrysene	Chr	$C_{18}H_{12}$	228	4	$8.4 \times 10^{-5*}$	2.0
Benzo(b)fluoranthene	BbF	$C_{20}H_{12}$	252	5	$6.7 \times 10^{-5*}$	0.80
Benzo(k)fluoranthene	BkF	$C_{20}H_{12}$	252	5	$1.3 \times 10^{-8*}$	0.76
Benzo(e)pyrene	BeP	$C_{20}H_{12}$	252	5	7.6×10^{-7}	6.3
Benzo(a)pyrene	BaP	$C_{20}H_{12}$	252	5	7.4×10^{-7}	3.8
Indeno(1,2,3-cd) pyrene	IP	$C_{22}H_{12}$	276	6	$1.3 \times 10^{-8*}$	62
Dibenz(a,h)anthracene	DahA	$C_{22}H_{14}$	278	5	$1.3 \times 10^{-8*}$	1.0
Benzo(g,h,i)perylene	BghiP	$C_{20}H_{12}$	276	6	1.4×10^{-8}	0.26

*Pa at 20°C

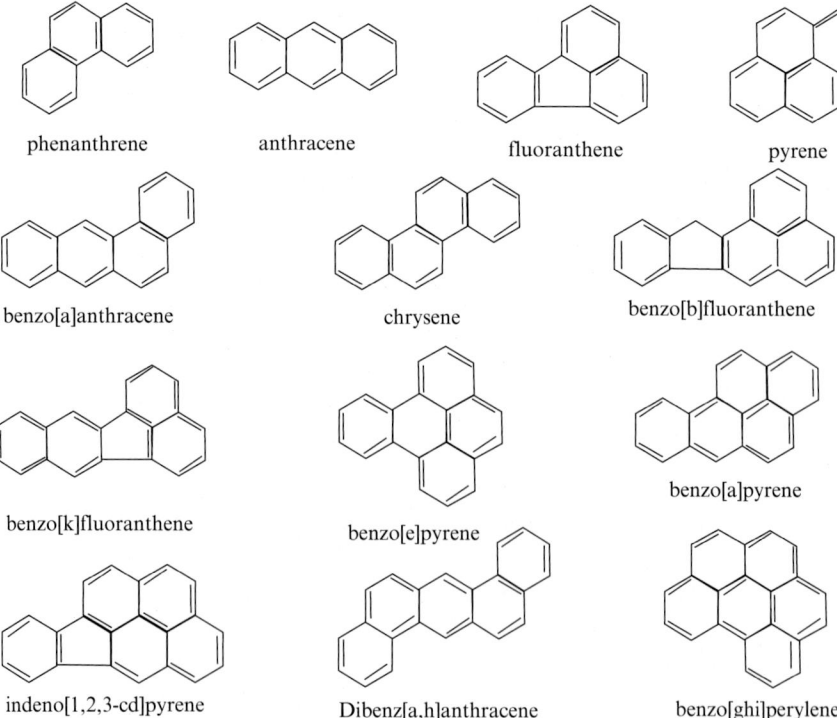

phenanthrene   anthracene   fluoranthene   pyrene

benzo[a]anthracene   chrysene   benzo[b]fluoranthene

benzo[k]fluoranthene   benzo[e]pyrene   benzo[a]pyrene

indeno[1,2,3-cd]pyrene   Dibenz[a,h]anthracene   benzo[ghi]perylene

Fig. 4.1 Structural formulas of 13 PAHs

standards for PAHs, and the United Kingdom has a proposed annual standard for BaP at 0.25 ng/m^3, though this value is often exceeded in both urban and rural areas, especially during the winter months (Lohmann et al. 2000). The Governmental Commission on Environmental Health (1996) of Sweden proposed environmental quality objectives for PAHs, including the goal for the long-term mean level of BaP not to exceed 0.1 ng/m^3 by 2020. All of the member states have used BaP as a marker, and Sweden has gone further and set a value for fluoranthene as well. Despite these trends of regulatory developments, countries may experience difficulties in demonstrating compliance because the source inventories for 1990 and current situations and our understanding of the atmospheric behavior of PAHs have major uncertainties (Prevedouros et al. 2004). However, the reduction of PAHs will have an indirect influence in that there is expected to be strengthened control of particulate material emissions from a wide range of sources, and PAH emissions are likely to be reduced together with particulate matter.

The atmospheric behaviors of PAHs include their transportation, dispersion, deposition, gas/particle distribution, particle size distribution, and reaction, which are largely influenced by meteorology. Traffic emissions are ground level, wide-spread, and concentrated in urban environments. The transportation, dispersion, and deposition of ambient PAHs, which are mostly associated with particulate matter, are closely related to the behavior of carrier particles, especially the size of the particles. Finer particles remain longer in the atmosphere due to less gravitational deposition. Additionally, a proportion of PAHs is subject to long-range atmospheric transport, which makes them a transboundary environmental problem. For example, Tamamura et al. (2007) reported a long-range transport of PAHs together with Asian dust from the eastern Asian continent to Kanazawa, Japan. During the transport, particulate PAHs experience various atmospheric processes, including gas/particle partitioning, and photochemical reactions. Those processes are reviewed in the following paragraphs.

Partitioning between the gas and particulate phases of PAHs is determined by temperature-dependent vapor pressure. Introducing subcooled liquid vapor pressure, Pankow and Bidleman (1991) demonstrated in actual field samples that the gas/particle partitioning of PAHs can effectively be described in the equation below:

$$LogK_p = m_r \log p°_L + b_r, \tag{4.1}$$

where K_p is the gas/particle partition coefficient and $p°_L$ is the subcooled liquid vapor pressure. The subcooled liquid vapor pressure is a function of temperature, where a higher temperature increases the pressure and consequently promotes the volatilization of PAHs from the particulate phase into the gas phase.

PAH species with molecular weights below that of pyrene exist mostly in the gas phase (e.g., Baker and Eisenreich 1990). Back et al. (1991) reported that PAHs with a molecular weight of less than 234 account for 50% or more of this group's total concentration during the summer and that these values decrease by about half during the winter. Gas-phase PAHs above or equal to a molecular weight of 252 constitute less than 20% of the total, even during the summer (Beak et al. 1991). On average, 47% of the total PAHs were reported as gas-phase PAHs (Beak et al. 1991). Back et al. (1991) reported that three-ring PAHs are predominantly gaseous, four-ring PAHs are a mixture of both phases, and five- and six-ring PAHs are primarily particulates. Pyrene was estimated to be 50% gaseous at 30°C by Westerholm et al. (1988). De Raat et al. (1990) reported that volatility was more important than the reaction in determining the PAH profile. For low molecular weight PAHs, especially those with weights up to that of chrysene, sampling effects may occur through possible volatilization and adsorption/desorption between gas/particle phases during the filter collection procedure of particles.

Concentrations of PAHs are found to be highly dependent upon the size of atmospheric aerosol, with the greatest concentrations being the submicron size range (e.g., Pierce and Katz 1975; Venkataraman et al. 1994, 1999; Venkataraman and Friedlander 1994). The higher concentrations at the submicron range can be explained by the condensation mechanism because larger specific surface areas are associated with such particles (e.g., Wiest and Fiorentina 1975). Size distribution is also explained by the Kelvin effect in several studies, which determined the relationship between the aerosol diameter and vapor pressure. As described in the equation below, more volatile species are associated with larger diameter particles:

$$\ln \frac{P}{P_0} = -\frac{2\gamma V_m}{rRT}, \tag{4.2}$$

where P: vapor pressure (Pa), P_0: saturated vapor pressure (Pa), R: gas constant (J/K·mol), T: temperature (K), γ: surface tension (N/m), V_m: molar volume (m^3/mol), and r: aerosol diameter (m). Consequently, low molecular weight PAHs often have bimodal distributions, while high molecular weight PAHs have unimodal distributions; these examples have been observed in numerous studies.

Atmospheric PAHs undergo various types of photochemical transformations upon reacting to OH radials, NO$_2$, ultraviolet rays, and so forth. Some of those derivatives are more toxic than others; in particular, nitro-PAHs are strong mutagens. Generally, photochemical transformations are the most important atmospheric decomposition processes of PAHs in both phases. Esteve et al. (2006) found that pyrene and benzo(a)pyrene are the most reactive with NO$_2$, whereas all PAHs studied presented similar reactivities with OH within a range of uncertainties. All PAHs appeared to be approximately four orders of magnitude more reactive with OH than with NO$_2$. These rate constants confirm that heterogeneous OH reactions are the dominant atmospheric loss process of PAHs compared with that of NO$_2$ reactions. The data are compared with that of previous literature reviews that describe both particulate and gas phases. This work demonstrates that the reactivity of PAHs in the gas phase is significantly higher than when associated with carbonaceous particulate substrates.

Ebert et al. (1988) reported the half-lives of PAHs under simulated sunlight (Table 4.2). The half-lives indicate that PAHs are quite reactive when exposed to oxygen in the presence of simulated sunlight. The photodecompositions of PAHs

Table 4.2 Half-lives of PAHs under simulated sunlight (IARC 2011)

PAH	Half-life (hours)
Ant	0.2
BaA	4.2
DahA	9.6
Pyr	4.2
BaP	5.3
BeP	21.1
BbF	8.7
BkF	14.1

Table 4.3 Half-lives of PAHs exposed to ozone (Ebert 1988)

PAH	Simulated sunlight plus 0.2 ppm O_3 (hours)	Dark reaction plus 0.2 ppm O_3 (hours)
Ant	0.15	1.23
BaA	1.35	2.88
DahA	4.80	2.71
Pyr	2.75	15.72
BaP	0.58	0.62
BeP	5.38	7.60
BbF	4.20	52.70
BkF	3.90	34.90

that exist on particles in a multilayered deposition show that the surface layer reacts very rapidly, exhibiting the reaction rate in Table 4.2, whereas the subsurface PAHs remain essentially protected from oxidation reactions. Various PAHs are also decomposed by nonphotochemical pathways, such as evaporative or oxidative reactions with gaseous pollutants. For example, the first kinetic studies of the heterogeneous decomposition of exposed PAHs in the unadsorbed state produced the high rate of reaction with ozone. The half-lives are shown in Table 4.3. The most rapid reactions were found for Ant, BaP, and BaA, with half-lives of 0.15, 0.58, and 1.35 h, respectively (Ebert 1988). This result implied that the dark reaction of several PAH toward O_3 underwent extremely fast reactions under simulated atmospheric conditions. As a result, significant degradation of PAHs can occur in O_3-polluted atmospheres. By contrast, in some experiments, no significant changes in the PAHs or mutagenic activity were observed for the ozonolysis of PAHs on diesel particulates (Ebert 1988).

The first order rate of the photochemical decomposition of PAHs was reported by Kamens et al. (1988). The photochemical decomposition occurred with PAH half-lives on the order of 1 h in field chamber tests under natural sunlight. Higher humidity and temperature and stronger sunlight promoted the decrease of PAH concentration. Among those meteorological factors, temperature was considered the least influential on the decay of PAHs. With regard to the atmospheric dilution effect of PAHs, the temporal change of mixing boundary layer height plays an important role both seasonally and diurnally. Height was estimated by the intensity of sunlight in a study conducted in Japan (Kim et al. 2001).

Table 4.4 Classification of the evaluation of carcinogenic risks to humans by the IARC (2011)

Evaluation of carcinogenic risks to humans	Group
Carcinogenic to humans	1
Probably carcinogenic to humans	2A
Possibly carcinogenic to humans	2B
Not classifiable as to its carcinogenicity to humans	3
Probably not carcinogenic to humans	4

Table 4.5 Classification of the evaluation of carcinogenic risks of individual PAHs by the IARC (2011)

PAH	Group
Phe	3
Ant	3
Fluo	3
Pyr	3
BaA	2B
Chr	2B
BbF	2B
BkF	2B
BeP	3
BaP	1
IP	2B
DahA	2A
BghiP	3

4.2.2 Risk Assessment

Carcinogenicity and mutagenicity are major human health concerns for PAHs. Several organizations, such as the International Agency for Research on Cancer (IARC), US EPA, EU, and the National Toxicology Program of the USA, have evaluated and classified the carcinogenic risks of the PAHs. According to the classification by the IARC (2011) (Table 4.4), carcinogenicity of individual PAHs analyzed in this study is indicated in Table 4.5. Quantitative cancer risk estimates have been attempted using in vitro and/or in vivo tests. For individual PAHs, cancer risk estimates for BaP are most abundant. Bostrom et al. (2002) summarized the BaP cancer risk estimates from animal experiments and epidemiological studies, and these are shown in Table 4.6. Among the unit risk values for BaP, 8.7×10^{-5} (per ng/m^3), as proposed by the WHO (1987, 2000), has been most frequently cited. The unit risk was estimated from epidemiological studies on coke oven workers in the USA and Canada. Another major unit risk value for BaP is 1.1×10^{-6} (per ng/m^3), which has been proposed by the California EPA. This unit risk was estimated from mouse experiments, which exposed particle-bound BaP to hamsters. The US EPA also initially proposed the same unit risk as the California EPA but then withdrew the value because of insufficient evidence for the estimate.

To assess the risk of exposure to a mixture of PAHs, a relative potency factor approach is generally taken. Benzo(a)pyrene is usually considered a surrogate, and

Table 4.6 Summary of unit risk estimates for BaP and for PAHs by BaP as the indicator substance (lifetime risk per ng/m^3 of BaP) (Bostrom et al. 2002)

Basis for calculation	Unit risk
Animal experiments	
Inhalation of BaP in hamsters (Saffiotti et al. 1972)	0.28×10^{-6} (RIVM 1989) $0.37–1.7 \times 10^{-6}$ (CARB 1994; Collins et al. 1991; Muller 1997)
Inhalation of BaP + SO$_2$ in rats (RIVM 1989)	0.59×10^{-6} (RIVM 1989)
Inhalation of BaP in mice (RIVM 1989)	400×10^{-6} (RIVM 1989)
Intratracheal instillation of BaP in hamsters (Saffiotti et al. 1972)	4.4×10^{-6} (CARB 1994; Collins et al. 1991)
Intratracheal instillation of BaP in hamsters (Feron et al. 1973)	4.8×10^{-6} (CARB 1994; Collins et al. 1991)
Inhalation of coal tar/pitch aerosol with BaP as the indicator substance	20×10^{-6} (Heinrich et al. 1994)
Epidemiology (PAH with BaP as indicator)	
US coke oven workers	87×10^{-6} (WHO 1987, 2000) 23×10^{-6} (Muller 1997) 50×10^{-6} (Pott 1985)
UK gas workers	430×10^{-6} (Pike 1983)
Smokey coal indoors in China	67×10^{-6} (RIVM 1989)
Most appropriate estimate	100×10^{-6} (RIVM 1989)
Aluminum smelters	90×10^{-6} (Armstrong et al. 1994)

Table 4.7 TEFs proposed for individual PAHs according to different authors

	Cal.EPA (2002)	Ontario (Muller) (1997)	Nisbet and LaGoy (1992)	Larsen and Larsen* (1998)
Phe	–	0.00064	0.001	0.0005
Ant	–	–	0.01	0.0005
Fluo	–	–	0.001	0.05
Pyr	–	0	0.001	0.001
BaA	0.1	0.014	0.1	0.005
Chr	0.01	0.026	0.01	0.03
BbF	0.1	0.11	0.1	0.1
BkF	0.1	0.037	0.1	0.05
BeP	–	0	–	0.002
BaP	1	1	1	1
IP	–	0.067	0.1	0.1
DahA	1.1	0.89	5	1.1
BghiP	–	0.012	0.01	0.02

*TEFs used in this study

toxicity equivalency quantity (TEQ) is evaluated by the sum of individual PAH doses multiplied by the toxicity equivalency factor (TEF). The currently proposed TEFs vary according to different authors (Table 4.7). Nisbet and LaGoy (1992) reviewed earlier relative potency estimates in 1992 and provided revised estimates.

The end points of the studies included carcinomas in the lungs of rats exposed via intrapulmonary administration, complete carcinogenesis in mouse skin, papillomas and/or carcinomas on mouse skin in initiation–promotion studies, sarcomas at the site of injection following subcutaneous administration to mice, and PAH–DNA adducts in in vitro studies. Relative potency factors (estimates of TEFs) were calculated using the data from each study by applying the same mathematical model of the dose–response relationship for each compound and comparing the results to those obtained for BaP. In a review of chemical carcinogens in the air, Larsen and Larsen (1998) listed the estimates of carcinogenic potencies of various PAHs relative to BaP. This TEF scheme is based on the extensive database on carcinogenicity studies using various routes of administration. In the table, some "noncarcinogenic PAHs" (fluoranthene, phenanthrene, pyrene) have been assigned TEFs such as 0.001, in contrast to the US EPA TEF scheme of 0. The authors claimed that assigning a TEF of 0.001 to the "noncarcinogenic PAHs" was motivated by their having "some, albeit limited, carcinogenic activity in some studies." However, this factor seems very uncertain, and even a low factor becomes important in cases with high levels. Kameda et al. (2005) calculated the geometrical averages of TEF values and applied these to a risk assessment study. In the present study, the TEFs documented by Larsen and Larsen (1998) were used because they were reported most recently, and values were available for all 13 PAHs measured in this study.

4.3 A Case Study on Field Measurement and Risk Assessment of PAHs in Bangkok, Thailand

Bangkok was selected for the case study. In this section, field measurement data of PAH concentrations are presented and utilized for risk assessment.

4.3.1 Field Measurement of PAH Concentrations

Bangkok is the capital of Thailand and has a population of more than 8 million people. Its climate is classified as tropical savanna with three seasons: hot (March to mid May), wet (mid May to October), and cool (November to February). The hot season is hot and dry, the wet season is hot and wet, and the cool season is cool and dry.

   Bangkok has a severe motor vehicle problem, with a large number of vehicles using diesel fuel. The number of vehicles has remarkably increased (Pollution Control Department and Thailand 2004), although there was a sudden drop in the vehicle number around 1997 due to the financial crisis. The increased number of vehicles is the direct cause of persistent road traffic congestion in Bangkok, which results in serious traffic air pollution. Currently, particulate matter is one of the main types of air pollutants, and it exceeds environmental standards. In particular, roadside PM10 levels consistently exceed the standard. Ruchirawat et al. (2002) reported that there are significant differences in the levels of exposure to atmospheric PAHs between traffic police and office duty police, with higher genotoxic effects for traffic police.

As background on the traffic air pollution in Bangkok, the dependency of transport and land use on the arterial roads is important (WHO 2000). Those arterial roads are poorly networked with a small road surface area of 2.4% (contrast this value with, for example, that of Tokyo, which is 16% or higher), where city buses serve as the primary public transport. In terms of land use, development is largely characterized by the promotion of arterial road-oriented dispersal of the population and commercial activities. An absolute lack of road space even in the core area has caused extreme congestion, which promotes the flight of development forces toward outer areas, leaving a large volume of unused or low-density land in the central area. Consequently, the absolute and relative number of roadside residents is quite high compared with that of other major cities, and thus, there is the concern for the risk of exposure to traffic air pollution. In Bangkok, diesel-fueled vehicles account for 28% of the in-use registered vehicle fleet, and these are estimated to emit 89% of the PM10 emission, while 1% is estimated to be emitted from gasoline-powered passenger cars and 10% from motorcycles (Asian Development 2006). Pickups, which are light-duty diesel vehicles, are major passenger cars in this country and account for about 30% of the passenger car share (Asian Development 2006).

4.3.1.1   Materials and Methods

Measurements of particulate phase PAHs were conducted on the roadside in Bangkok. Diurnal and seasonal variations of PAH concentrations were investigated by comparing two roadside sites with different road configurations. The measurement sites were Rama6 (R6) and Chockchai4 (CC). The R6 site was located in the area of government offices in the Bangkok city center, where one of the main roads, R6, carries heavy traffic. The R6 road is covered by an elevated highway (Fig. 4.2a), and this configuration, together with large roadside buildings, is likely to cause a stagnant air mass within the road space. By contrast, the CC site had an ordinary open-space configuration along the Ladphrao road, with many small shops and residential spaces on the upper floors (Fig. 4.2b). The measurement points were approximately 3 m from the roads at both sites, 1.5-m height from the ground at R6 (Fig. 4.3a) and 3-m height at CC in a Pollution Control Department's (PCD) air monitoring station, where the rooftop space of the station was provided to install measurement equipment for this study (Fig. 4.3b).

For air sampling, particle mass was collected using a 10-stage micro-orifice uniform-deposit impactor (MOUDI, Model 110, MSP Corporation, USA) (Corporation 1998) (Fig. 4.4). The principle operation of the MOUDI is the same as any inertial cascade impactor with multiple nozzles. At each stage, jets of particle-laden

Fig. 4.2 Measurement sites. (**a**) Rama6, (**b**) Chockchai4

Fig. 4.3 Equipment setup. (**a**) Rama6, (**b**) Chockchai4

Fig. 4.4 Micro-orifice uniform-deposit impactors (MOUDI)

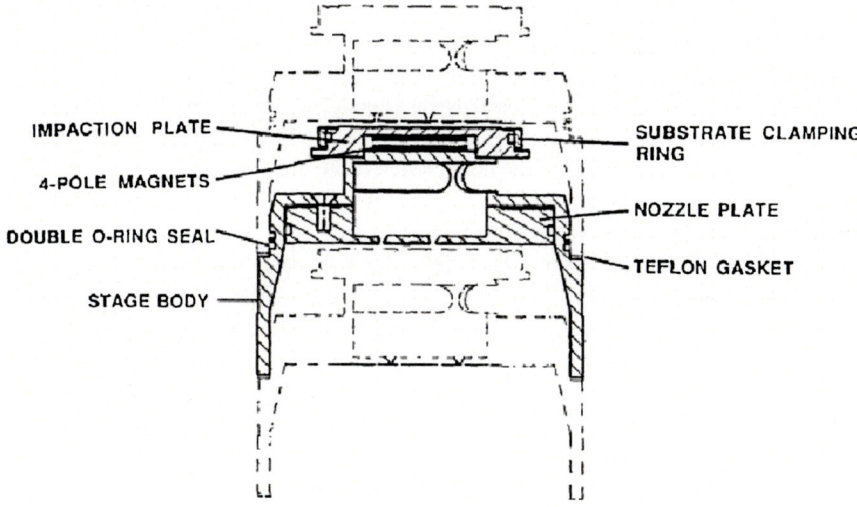

Fig. 4.5 Schematic diagram of a MOUDI stage showing its relation to the above and below stages (Saffiotti et al. 1972)

air impinge upon an impaction plate, and particles larger than the cut size of that stage cross the air streamlines and are collected upon the impaction plate. The smaller particles with less inertia do not cross the streamline and proceed to the next stage where the nozzles are smaller and where the air velocity through the nozzle is higher, and there, the finer particles are collected. This continues through the cascade impactor until the smallest particles are collected by the after-filter (Marple et al. 1991). Figure 4.5 shows a schematic diagram of one stage of the MOUDI, showing its relation to the above and below stages (Feron et al. 1973).

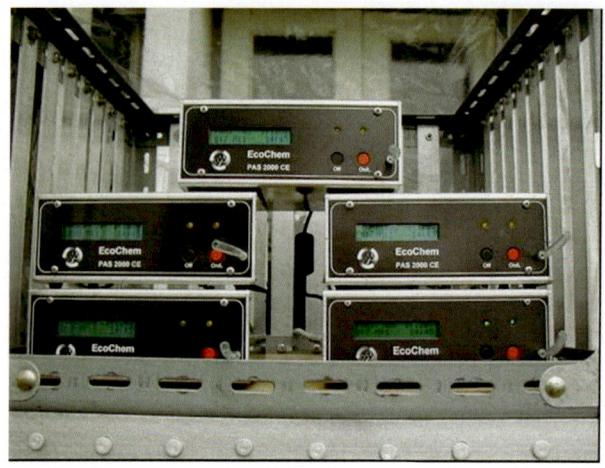

Fig. 4.6 PAS2000CE

Polytetrafluoroethylene (PTFE) membrane filters of 47-mm diameters (ADVANTEC, Japan) were used as the impaction substrates, and 37-mm glass filters (ADVANTEC, Japan) were used as the after-filter. The aerodynamic diameter size cut points with 50% collection efficiency were 0.18, 0.31, 0.56, 1.0, 1.8, 3.2, 5.6, 10, and 18 μm. By summation of the collected particle mass in the whole size range, the TSP concentrations were obtained. The MOUDI operated at 30 L/min, and the particle mass in the filters was determined gravimetrically. Before each weighing, the filters were conditioned in a desiccator with silica gel for about 3 days to eliminate humidity. Afterward, the filters were wrapped in aluminum foil and stored at 4°C until the extraction was performed. After ultrasonic extraction, the 13 PAHs with three to six aromatic rings, listed in Table 4.1, were determined by GC/MS analysis.

To monitor the temporal variations of particulate PAHs, photoelectric aerosol sensors (model PAS2000CE, EcoChem Analytics, Germany) (EcoChem 1999) were used for real-time monitoring (Fig. 4.6). Photoelectric aerosol sensors (PAS) work on the basis of photoelectric ionization of PAHs adsorbed onto particles (Burtscher and Schmidt-Ott 1986). The measurement techniques of this instrument have been described in detail elsewhere (Burtscher 1992). Briefly, a vacuum pump is used to draw ambient air through a quartz tube around which a UV lamp is mounted. Irradiation with UV light causes particles to emit electrons, which are then captured by surrounding gas molecules. Negatively charged particles are removed from the air stream, and the remaining positively charged particles are collected on a particle filter mounted in a Faraday cage. The particle filter converts the ion current to an electrical current, which is then amplified and measured with an electrometer (Fig. 4.7). The electric current establishes signals that are proportional to the concentrations of total PAHs (EcoChem 1999). The target particle size is below 1 μm.

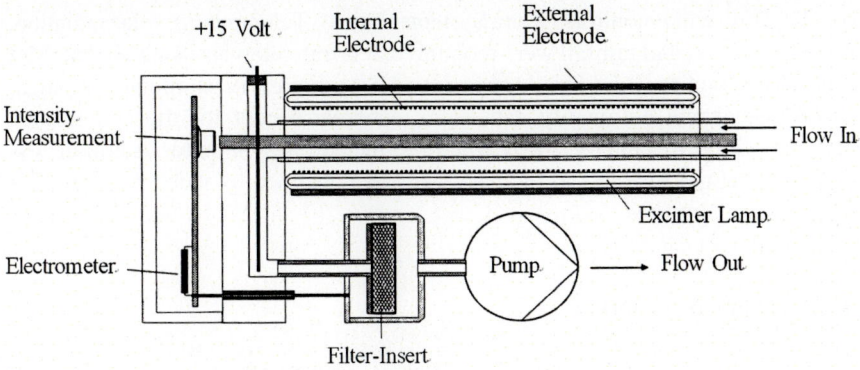

Fig. 4.7 Scheme of PAS2000CE (Marple et al. 1991)

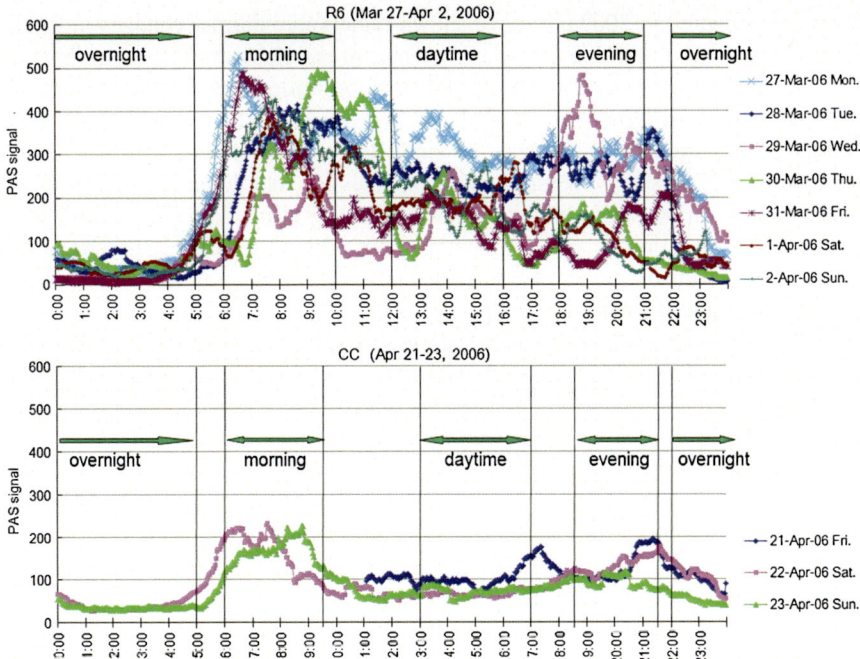

Fig. 4.8 Preliminary real-time monitoring for the selection of four time periods corresponding to peak and off-peak hours of PAS signals for MOUDI air sampling

At the CC site, hourly meteorology and air quality data monitored by the PCD were obtained. The meteorological data included temperature, solar radiation, relative humidity, rain, wind speed, and wind direction. The air quality data included NO_x, SO_x, ozone, PM10, and CO. At the R6 site, PM10 and CO were available. Also, wind speed and wind directions were monitored at 10-min intervals using KADEC wind monitors (Kona Systems, Japan). Temperature, solar radiation, relative humidity, and rainfall were monitored at 5-min intervals using an AutoMet meteorology monitor (Met One Instruments, USA). For every sampling campaign, meteorological and air quality data were available except for the cool season (December 2006 to January 2007) when the meteorology monitor AutoMet was out of order at the R6 site. Traffic data were also obtained by video recordings at both measurement sites.

4.3.1.2   Diurnal Variation

Size-fractioned PAH concentrations and their profiles are usually reported on a daily basis because measurements of particle size-fractioned PAHs using low-volume cascade air samplers are conducted over 1 day or longer periods in most cases. Some previous studies (e.g., Nielsen et al. 1996; Chetwittayachan 2002; Chetwittayachan et al. 2002a, b) have shown remarkable diurnal changes in PAH concentrations, with morning and evening peaks in parallel with traffic rush hours. Accordingly, there is interest in whether PAH concentrations and their profiles vary significantly over different time periods of day, especially between morning and evening peak periods and off-peak periods. Because significant photochemical degradation of PAHs may occur at different rates for individual types of PAHs, some as fast as on an hourly basis (Kamens et al. 1988; Brubaker and Hites 1998), it is likely that PAH profiles may vary within a day. This is particularly true because Thailand is a tropical country with strong sunlight. The diurnal variation further relates to the concern that the PAH profiles may vary to an extent that the actual PAH inhalation exposure varies significantly at any certain time period from the daily averaged values. However, such information has been scarce. Therefore, this subsection provides information on size-fractioned PAH profiles during different time periods of the day at the roadside in Bangkok.

   Particle size-fractioned PAH concentrations were measured using the MOUDI at the roadside during four different time periods of the day, namely, morning, daytime, evening, and overnight. Real-time monitoring of total PAH concentrations was also conducted using the PASs. A comparison was made between the two roadside sites, which had different building configurations along the roads (Fig. 4.2). Samples were collected in April 2006 during the hot season.

   Preliminary real-time monitoring of the continuous variations of PAH concentrations was conducted using the PASs from March 27 to April 2, 2006, at R6 and from April 21 to 23, 2006, at CC. Figure 4.8 presents the results from the PAS signals. Although the timing of the PAS signal peaks were somewhat varied on different days, the morning and evening peak hours and the daytime and overnight off-peak hours were estimated. Based on these observations, the following four time periods were selected for the MOUDI air sampling at each site: 6:00–10:00 (morning (m)), 12:00–16:00 (daytime (d)), 18:00–21:00 (evening (e)), and 22:00–5:00 (overnight (o)) at R6; and 6:00–9:30 (m), 13:00–17:00 (d), 18:30–21:30 (e), and 22:00–5:00 (o) at CC.

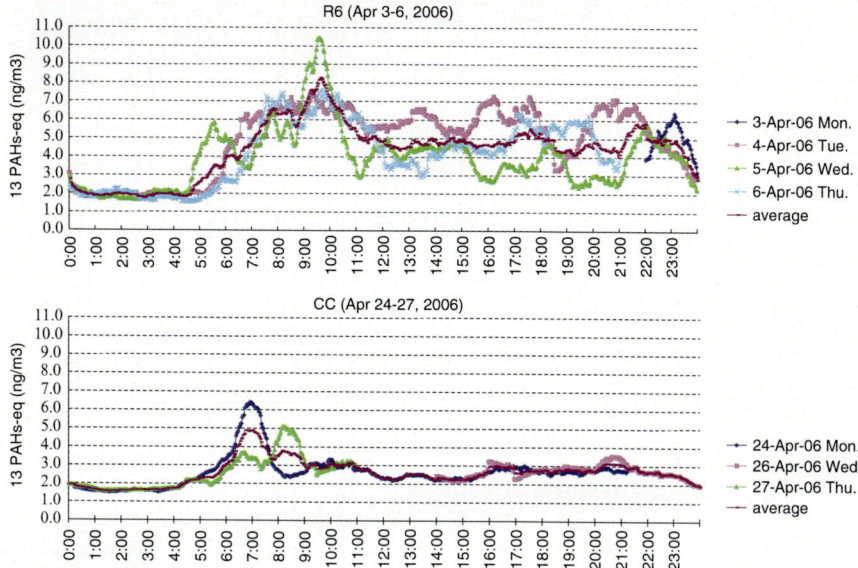

Fig. 4.9 Real-time monitoring of diurnal variations of 13 PAH equivalent concentrations (<1 μm)(ng/m³)

Air sampling was conducted using the MOUDI during the four selected time periods for three consecutive days. The sampling periods were April 3 (Mon.)–6 (Thu.), 2006, at R6 and April 24 (Mon.)–27 (Thu.), 2006, at CC. Particulate matter was collected cumulatively on the same filters in each time period, while the air sampling using the MOUDI was repeated for three consecutive days. After the sampling on the first and second days, the filters were kept in plastic cases and carried until the sampling on the second and third days. After the 3-day sampling, the sample filters were kept in a desiccator in a conditioned room of the PCD laboratory to eliminate humidity prior to particle mass weighing. Concurrent with the 3-day air sampling using the MOUDI, real-time monitoring was also conducted using the PASs. The 13 PAH concentrations obtained by the chemical analysis of MOUDI filter samples were confirmed to be sufficiently correlated with the PAS signals, and the PAS monitoring results are presented as equivalent 13 PAH concentrations in submicron particles converted from the PAS signals (Fig. 4.9).

At R6, the PAH concentration sharply increased from approximately 5 a.m. and reached morning peaks between 9 and 10 a.m. during the 3 days. Daytime concentrations were lower than in the morning. In the daytime and evening, several small peaks appeared. From around midnight to 5 a.m., concentrations were remarkably low. At CC, sharp morning peaks were observed at approximately 7 a.m. on April 24 and around 8 a.m. on April 27. In the evening, broader peaks appeared between 4 and 9 p.m., and then the concentration decreased. The sharp increase in the morning was observed at both sites. This observation is consistent with the observation by Chetwittayachan et al. (Chetwittayachan 2002;

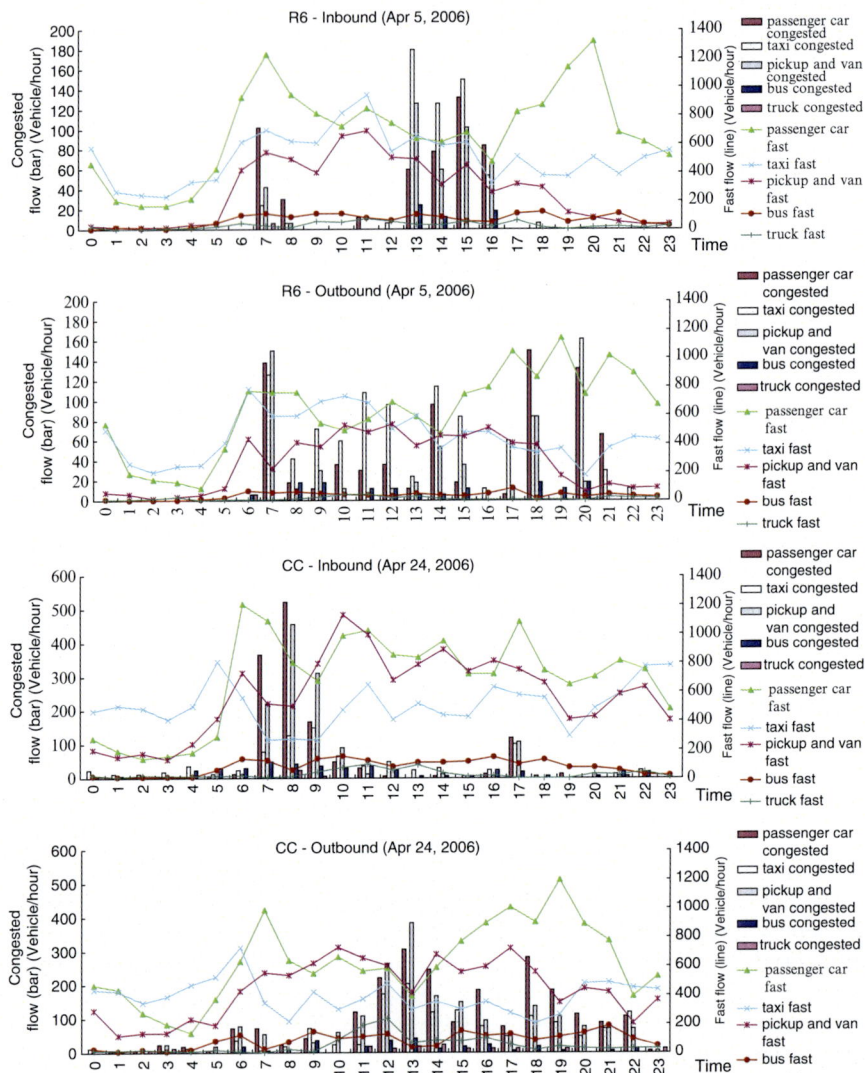

Fig. 4.10 Hourly traffic volume in fast and congested flows (April 2006)

Chetwittayachan et al. 2002a, b). The sharp morning peak can be explained by the strong atmospheric stability caused by the inversion layer and an increase in emissions from the morning traffic. Although the total traffic volume was smaller at R6 (74,000 vehicle/day on April 5 (Monday)) than that at CC (92,000 vehicle/day on April 24 (Wednesday)), higher concentrations were observed at R6 throughout the day, possibly due to the covered configuration that restricted the atmospheric dilution effect. Figure 4.10 shows the hourly traffic volume in fast and congested flow at R6 and CC. As shown in the figure, traffic congestion occurred during the

Table 4.8 Local meteorological data; average during the four time periods of the day (April 3–6, 2006, at R6 and April 24–27, 2006, at CC)

	Temperature (°C)				Solar radiation (W/m^2)				Relative humidity (%)			
	m	d	e	o	m	d	e	o	m	d	e	o
R6	29.5	35.9	31.2	29.7	26.1	284.8	0.2	0	85.3	57.9	76.3	83.8
CC	27.9	29.1	29.6	28.0	129.3	762.5	12.7	0.4	69.5	61.9	55.9	68.2

	Wind speed (m/s)				Wind direction			
R6	0.2	0.6	0.3	0.5	SW	WSW	SW	W
CC	1.9	1.3	1.8	2.8	SSW	SSW	SSW	SW

daytime at both sites, but daytime PAH concentrations at CC were constantly low compared with the large fluctuations of the daytime concentrations at R6.

The possible emission contribution from the elevated highway at the R6 sampling site is much smaller than that from the Rama6 road at the ground level because some previous dispersion studies (e.g., Jiang et al. 2004) report that emissions at a higher level are easily dispersed by faster air flow and by less surrounding obstacles than those at the ground level. The emission is unlikely to deposit directly below the highway.

The difference in road configurations also affected the local meteorological data at the two sites, especially wind speed and solar radiation. Table 4.8 shows the average local meteorological data during the 3-day monitoring periods. At both sites, the mean wind directions were almost stable at southwest, which meant that the wind flowed from the road to the sampling locations. At R6, the wind speed, which ranged from 0.2 to 0.6 m/s, was low compared with that at CC. The observation implies the limited dispersion and long residence time of PAHs at the site. On the other hand, the wind speed at CC was much higher, ranging 1.3–2.8 m/s, implying faster dispersion. Solar radiation, which promotes photochemical decomposition of PAHs (Kamens et al. 1988), was more than 40% lower at R6 throughout the day because of the elevated highway and the tall buildings along the R6 road. This elevated highway may also explain the higher daytime concentrations at R6.

The 13 particle size-fractioned PAH concentrations during the four time periods of the day were measured. The results shown in Table 4.9 are concentrations in the whole particle size range, which are summation of the size-fractioned concentrations. At R6, the 13 total PAH (13tPAH) concentrations in particles of the whole size range during the four time periods were 5.5 (m), 7.3 (d), 7.9 (e), and 3.3 ng/m^3 (o). At CC, the 13tPAH concentrations in particles of the whole size range during the four time periods were 4.1 (m), 4.1 (d), 4.0 (e), and 3.5 ng/m^3 (o). The concentrations observed at R6 were higher than that at CC, as expected from the results of the real-time monitoring. The contribution of the submicron particles was larger at R6 than at CC. At CC, from the PAS real-time monitoring results, the daytime concentration decreased quite sharply. By contrast, the 13tPAH concentration of the whole size range was 4.1 ng/m^3 in the morning, and it did not decrease in the daytime. However, the 13tPAH concentrations in the <1 μm range decreased from 3.3 ng/m^3 in the morning to 2.8 ng/m^3 in the daytime. This result seemed to contradict the real-time monitoring data, and the 13tPAH concentrations can be

Table 4.9 13 PAH concentrations (ng/m^3) in the whole particle size range

	R6				CC			
	m	d	e	o	m	d	e	o
Phe	0.31	0.51	0.50	0.20	0.46	0.48	0.51	0.45
Ant	0.15	0.37	0.35	0.12	0.19	0.23	0.24	0.28
Fluo	0.29	0.44	0.49	0.18	0.27	0.31	0.32	0.33
Pyr	0.50	0.73	0.73	0.30	0.38	0.42	0.46	0.41
BaA	0.33	0.51	0.55	0.19	0.21	0.31	0.26	0.24
Chr	0.40	0.64	0.60	0.23	0.34	0.30	0.27	0.35
BbF	0.64	0.83	0.89	0.25	0.46	0.56	0.45	0.42
BkF	0.40	0.58	0.69	0.31	0.28	0.29	0.22	0.25
BeP	0.65	0.58	0.58	0.33	0.45	0.40	0.41	0.25
BaP	0.74	0.65	0.89	0.42	0.38	0.21	0.35	0.24
IP	0.37	0.64	0.57	0.28	0.23	0.28	0.23	0.07
DahA	0.033	N.D.	0.13	0.017	N.D.	N.D.	N.D.	N.D.
BghiP	0.69	0.80	0.90	0.48	0.48	0.35	0.32	0.22
Σ13 PAHs	5.5	7.3	7.9	3.3	4.1	4.1	4.0	3.5

N.D. not detected

attributed to the fact that PASs detect PAHs that are associated with particles smaller than 1 μm. The PAH concentrations obtained in this study are remarkably lower than the PAH concentrations observed in 1996 by Garivait et al. (2001). On the other hand, comparable concentrations at a site near R6 were reported by Boonyatumanond et al. (2007) from measurements conducted in 2003. Concentrations of BaP, a typical marker of PAHs, ranged from 0.42 (o) to 0.89 ng/m^3 (e) at R6 to 0.21 (d) to 0.38 ng/m^3 (m) at CC. The BaP concentrations at R6 were approximately double the concentrations at CC.

4.3.1.3 Seasonal Variation

Although the PAH monitoring data of seasonal variations in temperate countries are relatively abundant, those in tropical countries are limited. There are some reports that particulate matter concentrations and associated PAH concentrations are lower during the wet season than the dry (hot or cool) season because of the deposition effects by tropical squalls (e.g., Panther et al. 1996). Those observational data, however, are not sufficient. Moreover, data on seasonal differences between hot and cool seasons are still scarce. This subsection presents comprehensive monitoring data of roadside particle size-fractioned PAH concentrations during all three seasons in Thailand. In addition to the seasonal variations of average PAH concentrations, those of diurnal variations of the PAH profiles are presented.

Air sampling was conducted using the MOUDI during the four time periods of the day and for continuous durations during 2 days in the three seasons at the R6 and CC sites. Concurrent real-time monitoring of PAHs by the PASs was also

Table 4.10 Measurement campaigns

			R6			CC	
			Air sampling			Air sampling	
Season	Period	Equipment	4 times periods of day	2 days		4 times periods of day	2 days
Wet	Sept. to Oct. 2005	MOUDI (PAS)		Oct. 18–20 (Oct. 17–25)		Sept. 26 to Oct. 7 (Sept. 26 to Oct. 7)	Sept. 13–15 (Sept.13–23)
Hot	Apr. 2006	MOUDI (PAS)	Apr. 3–6 (Apr. 3–10)	Apr. 10–12		Apr. 24–27 (Apr. 21–27)	Apr. 21–23
Cool	Dec. 2006 to Jan. 2007	MOUDI (PAS)	Dec. 26–29 (Dec. 21 to Jan. 15)	Dec. 21–22		Jan. 10–13 (Jan. 14–16)	Jan. 14–16

Table 4.11 Seasonal average of 13tPAH (<1 µm) equivalent concentrations and local meteorology during the real-time monitoring periods

	R6			CC		
	Hot	Cool	Wet	Hot	Cool	Wet
13 PAHs eq. (<1 µm) conc. (ng/m^3)	3.9	6.4	3.9	2.7	5.6	4.2
Temperature ($^\circ$C)	31.4	n.a.	29.0	29.0	29.3	26.6
Solar radiation (W/m^2)	80.7	n.a.	89.5	181.5	140.2	136.8
Rainfall (mm/h)	0.0	n.a.	0.0	0.0	0.0	1.0
Relative humidity (%)	76.7	n.a.	78.4	61.6	58.9	73.3
Wind speed (m/s)	0.32	n.a.	0.17	2.06	2.42	2.00
Wind direction (degree)	175	n.a.	143	224	224	254

n.a. data not available

conducted. The sampling procedure was the same as that of the previous subsection. The sampling campaigns are summarized in Table 4.10. In the table, the sampling campaign of the hot season in April 2006 was identical to that of the previous subsection. The diurnal sampling in the wet season was conducted on different days during the four time periods of the day. Thus, those data need to be considered with care in interpreting diurnal variation. The diurnal sampling in the hot and cool seasons was conducted on the same days during the four time periods so that their interpretation as diurnal change was possible.

In order to investigate seasonal variations, it is appropriate to discuss longer term monitoring data. In this respect, PAS monitoring results are discussed (Table 4.11). The table includes 13 total PAH equivalent concentrations in the submicron range, meteorology data of the corresponding periods. The seasonal averages of 13tPAH equivalent concentrations (<1 µm) were 3.9, 6.4, and 3.9 ng/m^3 for the hot, cool, and wet seasons at R6, respectively. Those at CC were 2.7, 5.6, and 4.2 ng/m^3. The seasonal average concentrations of 13tPAHs equivalent concentrations were highest in the cool season at both sites. The main reason should be the cooler climate than the other two seasons, which led to lower mixing layer height, consequently reduced atmospheric dilution. In a comparison between R6 and CC, the concentrations at R6 exhibited higher values in the hot and cool seasons but not in the wet season. The reason for the lower concentration at R6 in the wet season could be attributed to the local wind direction and transport of road emissions, which means in the wet season, wind was blowing along the road or from the sampling location to the road at R6, whereas at CC wind was blowing from the road, transporting road emissions to the sampling location.

4.3.2   Risk Assessment of PAHs for Roadside Residents

The US EPA states that the lifetime risk of 1.0×10^{-5} is a critical level considered for significant adverse health effects. In the case of airborne PAHs, lung cancer is the primary possible adverse effect. Conventional risk assessment of PAHs has used the daily average or annual average PAH concentrations. However, this study obtained more accurate PAH risk assessment, with consideration of diurnal and seasonal concentration variations, by utilizing the measurement data of size-fractioned 13tPAHs. The data included 13 PAH concentrations during four time periods of the day during three seasons at the R6 and CC sites. There are basically 12 datasets of PAH profiles available to estimate the lifetime risk at each site. For R6, only 2 days of average data are available for the wet season instead of the diurnal variation data; thus, there are nine datasets available. The relative contributions to the lifetime risk were estimated. This method allows for the evaluation of PAH risk for roadside residents who have different lifestyles in terms of their exposure time periods of the day. Furthermore, seasonal variation data improved the accuracy of estimating the lifetime risk.

4.3.2.1   Calculation of Lifetime Cancer Risk

In the assessment of the cancer risk due to PAHs in humans, the unit risk is usually applied. The unit risk value for BaP (ng/m3) 8.7×10^{-5} (WHO 1987, 2000) was adopted in this study. The relative potency factor approach is commonly taken in assessing health risk due to a mixture of PAHs. This approach utilizes toxicity equivalency factors (TEF), which refer to a representative toxicity of a surrogate compound, usually BaP. Those values are proposed from several organizations or researchers based on in vitro or in vivo experiments using animals. The values vary according to different sources, as mentioned in Sect. 4.2.2 (Table 4.7), often due to differences in the methodologies of the experiments. In this study, the TEFs documented by Larsen and Larsen (1998) were applied because the TEF values were available for all of the PAHs (Table 4.7). The TEFs were multiplied by the dose, namely, individual PAH concentrations (ng/m^3), then summed to make a BaP-toxicity equivalency quantity (TEQ) for a mixture of PAHs. By multiplying the unit risk value of 8.7×10^{-5}, the risk of the mixture was assessed.

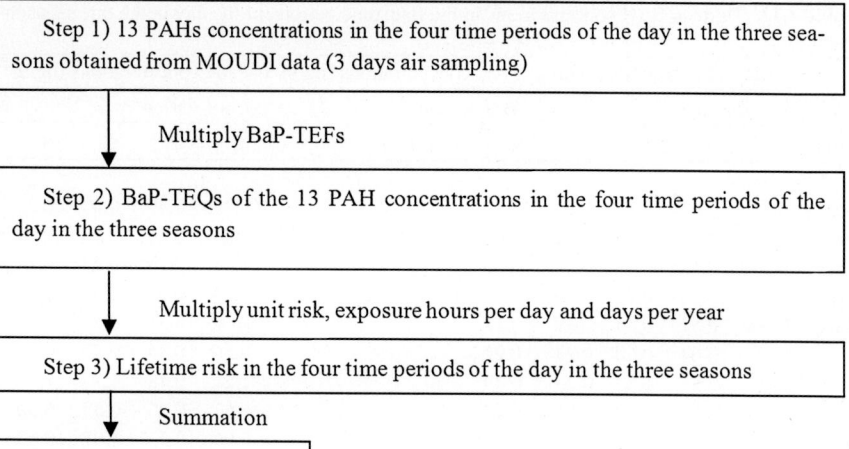

Fig. 4.11 Calculation steps of lifetime risk

Lifetime PAH cancer risk was estimated by calculating the lifetime risk from exposure during each time period of the day during each season. The MOUDI air sampling durations of the morning, daytime, evening, and overnight periods were extended to 6 h for the calculation of exposure duration, which included morning 5:00–11:00 (m), daytime 11:00–17:00 (d), evening 17:00–23:00 (e), and overnight 23:00–5:00 (o). Equation 4.3 shows the risk calculation. Figure 4.11 shows the calculation steps. The calculation results are shown according to these steps. Furthermore, estimating the PAH lifetime cancer risk considers the contributions of individual PAHs and different particle size fractions:

$$Cancer\ risk = unit\ risk \times \overset{3\ seasons}{\underset{k=1}{\sum}}\ \overset{4\ time\ periods\ of\ the\ day}{\underset{j=1}{\sum}}\ \overset{13\ PAHs}{\underset{i=1}{\sum}} (S/365)(D/24)C_{i,j,k} \times TEF_i,$$

$$(4.3)$$

where:

i: type of PAH ($i = 1$–13)
j: time period of the day ($j = 1$–4)
k: season of the year ($k = 1$–3)
C: PAH concentration (ng/m^3)
D: hours in each time period of the day ($D = 6$ h)
S: days in each season of the year ($S = 76$, 169, and 120 days in the hot, wet, and cool seasons, respectively)
$unit\ risk$: 8.7×10^{-5} (ng/m^3)$^{-1}$

Thirteen individual PAH concentrations of the whole particle size range used for the risk calculations are shown in Table 4.12 (step 1). BaP-TEQs were calculated by

Table 4.12 Thirteen PAH concentrations in the four time periods of the day in the three seasons (Step 1)

PAH	PAH concentrations (ng/m^3) Hot season				PAH concentrations (ng/m^3) Wet season 2-day average	PAH concentrations (ng/m^3) Cool season			
	m	d	e	o		m	d	e	o
(a) R6									
Phe	0.31	0.51	0.50	0.20	0.13	0.47	0.30	0.40	0.23
Ant	0.15	0.37	0.35	0.12	0.066	0.23	0.14	0.16	0.09
Fluo	0.29	0.44	0.49	0.18	0.13	0.44	0.26	0.34	0.24
Pyr	0.50	0.73	0.73	0.30	0.20	0.54	0.39	0.51	0.31
BaA	0.33	0.51	0.55	0.19	0.18	0.46	0.31	0.34	0.22
Chr	0.40	0.64	0.60	0.23	0.22	0.56	0.34	0.57	0.30
BbF	0.64	0.83	0.89	0.25	0.44	1.2	0.48	1.1	1.2
BkF	0.40	0.58	0.69	0.31	0.23	0.57	0.26	0.57	0.50
BeP	0.65	0.58	0.58	0.33	0.37	0.96	0.52	1.07	0.93
BaP	0.74	0.65	0.89	0.42	0.26	0.67	0.40	0.65	0.68
IP	0.37	0.64	0.57	0.28	0.31	0.96	0.58	0.87	0.80
DahA	0.033	N.D.	0.13	0.017	N.D.	N.D.	N.D.	N.D.	0.071
BghiP	0.69	0.80	0.90	0.48	0.45	1.3	0.82	1.5	1.3
Total	5.5	7.3	7.9	3.3	3.0	8.3	4.8	8.1	6.9

PAH	PAH concentrations (ng/m^3) Hot season				PAH concentrations (ng/m^3) Wet season				PAH concentrations (ng/m^3) Cool season			
	m	d	e	o	m	d	e	o	m	d	e	o
(b) CC												
Phe	0.46	0.48	0.51	0.45	0.69	0.69	0.61	0.45	0.70	1.2	0.46	0.35
Ant	0.19	0.23	0.24	0.28	0.54	0.58	0.30	0.25	0.32	0.44	0.18	0.20
Fluo	0.27	0.31	0.32	0.33	0.59	0.60	0.44	0.36	0.54	0.90	0.38	0.35
Pyr	0.38	0.42	0.46	0.41	0.63	0.66	0.43	0.36	0.68	0.95	0.48	0.40
BaA	0.21	0.31	0.26	0.24	0.28	0.31	0.29	0.24	0.33	0.32	0.22	0.22
Chr	0.34	0.30	0.27	0.35	0.30	0.31	0.36	0.26	0.65	0.53	0.42	0.30
BbF	0.46	0.56	0.45	0.42	0.56	0.40	0.42	0.95	1.23	0.57	0.45	1.16
BkF	0.28	0.29	0.22	0.25	0.35	0.28	0.42	0.47	0.98	0.69	0.64	0.59
BeP	0.45	0.40	0.41	0.25	0.45	0.26	0.30	0.58	1.11	0.73	0.67	0.83
BaP	0.38	0.21	0.35	0.24	0.48	0.27	0.38	0.89	0.91	0.15	0.39	0.85
IP	0.23	0.28	0.23	0.07	0.70	0.41	0.53	1.1	0.74	0.96	0.65	0.84
DahA	N.D.	N.D.	N.D.	N.D.	0.020	N.D.	0.15	0.13	0.017	0.059	N.D.	0.094
BghiP	0.48	0.35	0.32	0.22	1.0	0.64	0.87	1.6	1.1	0.76	1.1	1.1
Total	4.1	4.1	4.0	3.5	6.6	5.4	5.5	7.6	9.3	8.3	6.0	7.2

N.D. not detected

multiplying the TEFs, as shown in Table 4.13 (step 2). Table 4.14 (step 3) shows the lifetime risk in the four diurnal time periods in the three seasons. Table 4.15 (step 4) summarizes the lifetime risk at respective sites.

Lifetime risk was estimated to be 5.6×10^{-5} for R6 and 6.5×10^{-5} for CC. As shown in Table 4.15, the contributions were 29.2%, 26.1%, and 44.7% for the hot, wet, and cool seasons, respectively, at R6 and 11.5%, 49.2%, 39.3%, respectively, at CC. The lifetime risk was higher at CC than at R6. The PAH concentrations in the

Table 4.13 BaP-TEQs in the four time periods of the day in the three seasons (Step 2)

| | BaP-TEQ (ng/m³) Hot season | | | | BaP-TEQ (ng/m³) Wet season | BaP-TEQ (ng/m³) Cool season | | | |
	m	d	e	o	2-day average	m	d	e	o
PAH									
(a) R6									
Phe	0.00015	0.00026	0.00025	0.00010	0.000067	0.00023	0.00015	0.00020	0.00012
Ant	0.000076	0.00018	0.00017	0.000061	0.000033	0.00011	0.000068	0.000080	0.000046
Fluo	0.014	0.022	0.025	0.0088	0.0064	0.022	0.013	0.017	0.012
Pyr	0.00050	0.00073	0.00073	0.00030	0.00020	0.00054	0.00039	0.00051	0.00031
BaA	0.0016	0.0026	0.0028	0.0010	0.00089	0.0023	0.0016	0.0017	0.0011
Chr	0.012	0.019	0.018	0.007	0.0065	0.017	0.010	0.017	0.009
BbF	0.064	0.083	0.089	0.025	0.044	0.12	0.048	0.11	0.12
BkF	0.020	0.029	0.034	0.015	0.011	0.029	0.013	0.028	0.025
BeP	0.0013	0.0012	0.0012	0.00066	0.00073	0.0019	0.0010	0.0021	0.0019
BaP	0.74	0.65	0.89	0.42	0.26	0.67	0.40	0.65	0.68
IP	0.037	0.064	0.057	0.028	0.031	0.096	0.058	0.087	0.080
DahA	0.036	0.0	0.142	0.019	0.0	0.0	0.0	0.0	0.079
BghiP	0.014	0.016	0.018	0.010	0.0091	0.025	0.016	0.031	0.025
Total	0.94	0.88	1.3	0.54	0.37	1.0	0.56	0.94	1.0

(continued)

Table 4.13 (continued)

PAH	BaP-TEQ (ng/m³) Hot season				BaP-TEQ (ng/m³) Wet season				BaP-TEQ (ng/m³) Cool season			
	m	d	e	o	m	d	e	o	m	d	e	o
(b) CC												
Phe	0.00023	0.00024	0.00025	0.00023	0.00034	0.00034	0.00031	0.00022	0.00035	0.00062	0.00023	0.00017
Ant	0.00009	0.00011	0.00012	0.00014	0.00027	0.00029	0.00015	0.00013	0.00016	0.00022	0.000091	0.00010
Fluo	0.013	0.015	0.016	0.017	0.029	0.030	0.022	0.018	0.027	0.045	0.019	0.017
Pyr	0.00038	0.00042	0.00046	0.00041	0.00063	0.00066	0.00043	0.00036	0.00068	0.00095	0.00048	0.00040
BaA	0.0011	0.0015	0.0013	0.0012	0.0014	0.0016	0.0015	0.0012	0.0017	0.0016	0.0011	0.0011
Chr	0.010	0.0089	0.0081	0.010	0.0090	0.0093	0.011	0.0078	0.019	0.016	0.013	0.009
BbF	0.046	0.056	0.045	0.042	0.056	0.040	0.042	0.095	0.123	0.057	0.045	0.116
BkF	0.014	0.014	0.011	0.013	0.018	0.014	0.021	0.024	0.049	0.034	0.032	0.029
BeP	0.0009	0.0008	0.0008	0.0005	0.0009	0.0005	0.0006	0.0012	0.0022	0.0015	0.0013	0.0017
BaP	0.38	0.21	0.35	0.24	0.48	0.27	0.38	0.89	0.91	0.15	0.39	0.85
IP	0.023	0.028	0.023	0.007	0.070	0.041	0.053	0.111	0.074	0.096	0.065	0.084
DahA	0.0	0.0	0.0	0.0	0.022	0.00	0.16	0.14	0.018	0.064	0.0	0.104
BghiP	0.010	0.0070	0.0064	0.0044	0.021	0.013	0.017	0.031	0.022	0.015	0.022	0.021
Total	0.50	0.35	0.46	0.33	0.71	0.42	0.71	1.3	1.2	0.49	0.58	1.2

Table 4.14 Lifetime risk in the four diurnal time periods in the three seasons and their contribution (%) to the lifetime risk (Step 3)

	Hot season		Wet season		Cool season		Total of the three seasons	
	Risk	%	Risk	%	Risk	%	Risk	%
(a) R6								
(m) 5:00–11:00	4.3×10^{-6}	7.6	1.5×10^{-5} (24 h)	26.1	7.0×10^{-6}	12.5		
(d) 11:00–17:00	4.0×10^{-6}	7.1			4.0×10^{-6}	7.1		
(e) 17:00–23:00	5.8×10^{-6}	10.2			6.7×10^{-6}	11.9		
(o) 23:00–5:00	2.4×10^{-6}	4.3			7.4×10^{-6}	13.2		
Total	1.6×10^{-5}	29.2	1.5×10^{-5}		2.5×10^{-5}	44.7		
(b) CC								
(m) 5:00–11:00	2.3×10^{-6}	3.5	7.2×10^{-6}	11.1	8.9×10^{-6}	13.8	1.8×10^{-5}	28.4
(d) 11:00–17:00	1.6×10^{-6}	2.4	4.2×10^{-6}	6.6	3.5×10^{-6}	5.4	9.3×10^{-6}	14.4
(e) 17:00–23:00	2.1×10^{-6}	3.2	7.1×10^{-6}	11.0	4.2×10^{-6}	6.5	1.3×10^{-5}	20.7
(o) 23:00–5:00	1.5×10^{-6}	2.3	1.3×10^{-5}	20.6	8.8×10^{-6}	13.7	2.4×10^{-5}	36.6
Total	7.5×10^{-6}	11.5	3.2×10^{-5}	49.2	2.5×10^{-5}	39.3	6.5×10^{-5}	100

Table 4.15 Lifetime risk (Step 4)

R6	CC
5.6×10^{-5}	6.5×10^{-5}

four time periods of the day during the 3-day air sampling showed lower concentrations at R6 in the wet and the cool seasons, in contrast to results obtained by real-time monitoring and the 2-day continuous air sampling. On average, the seasonal average concentrations showed higher concentrations at R6. The possible reasons for this observation are that during the 3-day air sampling period in the wet season, the wind was blowing from the sampling site to the road or along the road at R6 and that the sampling period in the cool season at R6 was the end of the year when the traffic was possibly lighter. If the sampling periods had been longer, the seasonal average concentrations would have been higher at R6.

As shown in Table 4.15, the lifetime risk in an individual season was already higher than 1.0×10^{-5} at both sites, except for the hot season at CC. The contributions of each time period of the day to the lifetime risk were also estimated for CC: 28.4% in the morning, 14.4% in the daytime, 20.7% in the evening, and 36.6% in the overnight. The contributions of the morning and overnight risks were much higher than those of the daytime and the evening. The higher contributions of BaP in the morning and overnight periods were apparent, especially in the wet and cool seasons, as shown in Table 4.13b. The BaP concentrations in the daytime at CC were observed to be lower and were probably subjected to photochemical degradation.

The contributions of risk per hour to the estimated lifetime risk are shown in Table 4.16. The hourly contributions of the different time periods of the day and the

Table 4.16 Contributions of risk per hour to the estimated lifetime risk ($10^{-2}\%$)

	Hot season	Wet season	Cool season	Total of the three seasons
(a) R6				
(m) 5:00–11:00	1.7	0.64	1.7	
(d) 11:00–17:00	1.6		1.0	
(e) 17:00–23:00	2.2		1.7	
(o) 23:00–5:00	0.95		1.8	
Average	1.6	0.64	1.6	
(b) CC				
(m) 5:00–11:00	0.77	1.1	1.9	1.3
(d) 11:00–17:00	0.53	0.65	0.75	0.66
(e) 17:00–23:00	0.70	1.1	0.90	0.95
(o) 23:00–5:00	0.50	2.0	1.9	1.7
Average	0.63	1.2	1.4	1.1

Table 4.17 Exposure duration (hour) in each time period of the day at CC which leads to the lifetime risk of 1.0×10^{-5}

	Morning	Daytime	Evening	Overnight
Hour	3.3	6.5	4.5	2.5

different seasons ranged between 0.64 and $2.2 \times 10^{-2}\%$ at R6 and 0.53 and $1.9 \times 10^{-2}\%$ at CC. The differences in the contributions were up to 3.6 times at R6 comparing the wet season and the evening period in the hot season and 4.1 times at CC comparing the overnight period in the wet season and the overnight period in the hot season. Consequently, the significance of the risk depends considerably on the exposure time periods. Table 4.17 shows the calculated daily exposure durations in each time period of the day at CC, which may lead to the lifetime risk of 1.0×10^{-5}. The daily 2.5 h of exposure during the overnight period was the highest level of exposure for the lifetime at CC, leading to the lifetime risk of 1.0×10^{-5}.

Gamo et al. (1996) propose an estimation method for the loss of life expectancy (LLE) from cancer risk. In this study, the LLE was introduced to show the adverse effects of PAHs in a more dynamic way in addition to the lifetime cancer risk assessment. The LLE estimation took advantage of the PAH measurement data in the different time periods of the day in the different seasons. In the LLE estimation model, the cancer risk of 1.0×10^{-5} corresponded to an LLE of 65.8 min. Table 4.18 shows the LLEs from the lifetime exposure to the 13 individual PAHs in the four time periods of the day in the three seasons. The total LLEs were approximately 6.1 h at R6 and 7.1 h at CC.

4.3.2.2 Contributions of Different Types of PAHs and Particle Sizes

The lifetime risks of individual PAHs in the four diurnal time periods in the three seasons are shown in Fig. 4.12, and those for the whole year are shown in Fig. 4.13.

Table 4.18 LLE (minutes) from the lifetime exposure to the 13 individual PAHs in the four time periods of the day in the three seasons

PAH	LLE (min)				LLE (min)	LLE (min)				LLE (min)
	m	d	e	o	2-day average	m	d	e	o	Yearly total
(a) R6										
Phe	0.00	0.01	0.01	0.00	0.00	0.01	0.01	0.01	0.01	0.06
Ant	0.00	0.01	0.01	0.00	0.00	0.01	0.00	0.00	0.00	0.03
Fluo	0.43	0.66	0.74	0.26	0.43	1.04	0.62	0.80	0.56	5.54
Pyr	0.01	0.02	0.02	0.01	0.01	0.03	0.02	0.02	0.01	0.16
BaA	0.05	0.08	0.08	0.03	0.06	0.11	0.07	0.08	0.05	0.61
Chr	0.36	0.57	0.54	0.21	0.43	0.79	0.48	0.81	0.43	4.62
BbF	1.92	2.47	2.65	0.75	2.93	5.55	2.28	5.22	5.85	29.6
BkF	0.60	0.87	1.03	0.46	0.76	1.35	0.60	1.34	1.19	8.21
BeP	0.04	0.03	0.03	0.02	0.05	0.09	0.05	0.10	0.09	0.51
BaP	22.1	19.4	26.5	12.7	17.0	31.8	18.7	30.6	32.3	211
IP	1.11	1.91	1.72	0.85	2.09	4.56	2.75	4.09	3.78	22.9
DahA	1.09	0.00	4.24	0.57	0.00	0.00	0.00	0.00	3.72	9.62
BghiP	0.41	0.48	0.54	0.29	0.61	1.19	0.78	1.44	1.18	6.91
13tPAHs	28.2	26.5	38.1	16.1	97.5	46.5	26.4	44.5	49.2	373
Seasonal total	109				97.5	167				373

PAH	LLE (min)				LLE (min)				LLE (min)				LLE (min)
	m	d	e	o	m	d	e	o	m	d	e	o	Yearly total
(b) CC													
Phe	0.01	0.01	0.01	0.01	0.02	0.02	0.02	0.01	0.02	0.03	0.01	0.01	0.17
Ant	0.00	0.00	0.00	0.00	0.02	0.02	0.01	0.01	0.01	0.01	0.00	0.00	0.10
Fluo	0.40	0.46	0.47	0.50	1.96	1.99	1.48	1.20	1.27	2.13	0.90	0.82	13.58
Pyr	0.01	0.01	0.01	0.01	0.04	0.04	0.03	0.02	0.03	0.04	0.02	0.02	0.31
BaA	0.03	0.05	0.04	0.04	0.09	0.10	0.10	0.08	0.08	0.08	0.05	0.05	0.78
Chr	0.31	0.27	0.24	0.31	0.60	0.62	0.71	0.52	0.92	0.75	0.59	0.42	6.27
BbF	1.37	1.67	1.36	1.25	3.73	2.69	2.76	6.33	5.81	2.72	2.11	5.47	37.3
BkF	0.42	0.43	0.33	0.38	1.18	0.93	1.40	1.58	2.31	1.63	1.51	1.39	13.5
BeP	0.03	0.02	0.02	0.02	0.06	0.03	0.04	0.08	0.10	0.07	0.06	0.08	0.62
BaP	11.5	6.42	10.5	7.12	32.3	18.0	25.1	59.5	43.1	7.3	18.3	40.3	279
IP	0.69	0.83	0.68	0.22	4.65	2.76	3.53	7.38	3.51	4.53	3.08	3.99	35.9
DahA	0.00	0.00	0.00	0.00	1.44	0.00	10.8	9.29	0.87	3.05	0.00	4.91	30.3
BghiP	0.29	0.21	0.19	0.13	1.39	0.86	1.16	2.09	1.02	0.72	1.03	1.00	10.1
13tPAH	15.1	10.4	13.9	10.0	47.5	28.1	47.1	88.1	59.1	23.0	27.6	58.5	428
Seasonal total	49.3				211				168				428

m morning (5:00–11:00), *d* daytime (11:00–17:00), *e* evening (17:00–23:00), *o* overnight (23:00–5:00)

The contributions of BaP were predominant in all of the time periods. From Fig. 4.13, the BaP contributions were 57% at R6 and 64% at CC. Besides BaP, the contributions of BbF, IP, and DahA were significant, although the contributions of individual PAHs were considerably dependent on the selection of TEF values.

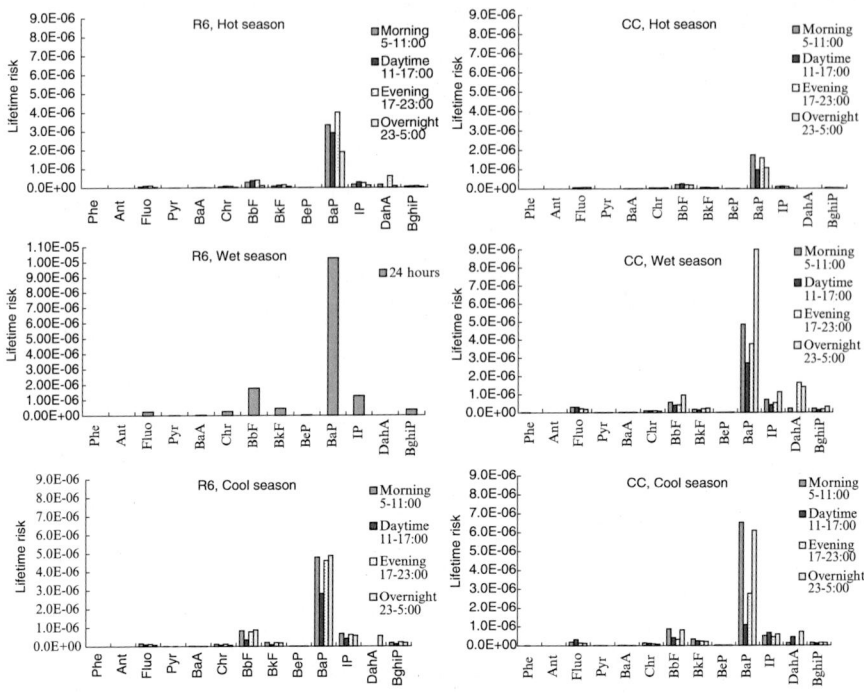

Fig. 4.12 Lifetime risk of individual PAHs in the four time periods of the day in the three seasons

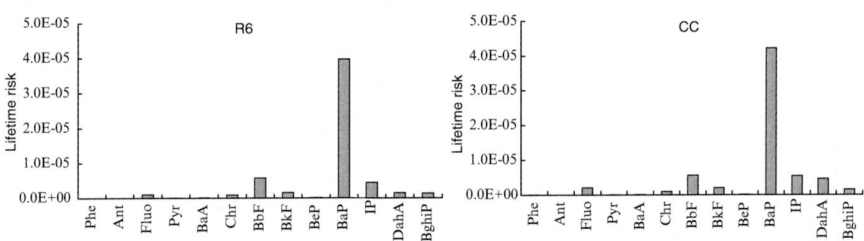

Fig. 4.13 Contribution of the individual PAHs to the lifetime risk

The lifetime risks according to the different particle size fractions were estimated. The contributions of PAHs in the PM0.18, PM1.8, and PM10 fractions are shown in Fig. 4.14 for each time period of the day in each season. The table shows that most of PAH risk can be attributed to finer particle size fractions than 1.8 μm.

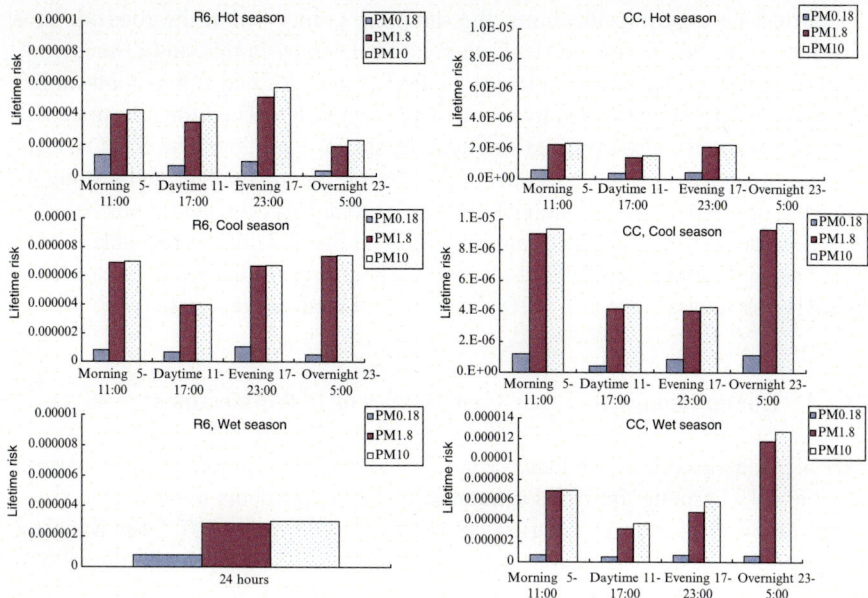

Fig. 4.14 Lifetime risk according to the different particle size fractions in the four time periods of the day in the three seasons

4.3.2.3   Uncertainty of Risk Assessment

There are several factors of uncertainty in risk assessment, such as:

- Recovery rates in the chemical analysis procedure
- GC/MS detection limits
- Measurement location
- Types of equipment
- Risk assessment method, especially unit risk values

When the recovery of the chemical analysis procedure was corrected, the concentrations of Ant, IP, DahA, and BghiP were mainly affected. Ant had a too-high recovery (206%), and IP, DahA, and BghiP had relatively low recovery rates (around 70%), which were underestimated in the last section. When the recovery rates were corrected, the 13tPAH concentrations increased from 4% to 15%. The BaP-TEQs increased from 9% to 19%. The increase of BaP-TEQs was larger than that of the concentrations because the underestimated PAHs, namely, IP, DahA, and BghiP, were given relatively high TEFs. The risk was estimated at 6.3×10^{-5} at R6 and 7.3×10^{-5} at CC. These values increased by 11% at R6 and 11% at CC from the risk estimate values without the recovery rate correction, although the percentages of contributions of each time period did not change significantly.

Considering exposure locations, the sampling points within the roadside area must have affected the risk levels. Using one field survey in this study framework, we investigated the spatial variation of the PAS signals located at five points at R6. The average PAS signal varied by as much as a factor of 9. Therefore, accurate risk assessments should also consider this type of spatial variation. The PAH risk for nonroadside residents is also concerning. According to another field study in Bangkok, differences in the 1-month average 13 total PAH concentrations between general areas and the roadside were only 1.6–3.0 times, with the roadside having higher concentrations (Hoshiko et al. 2005). Therefore, for nonroadside residents, the PAH risk may exceed 1.0×10^{-5} as well and is not a negligible level.

4.3.2.4 Comparison with Other Case Studies of Risk Assessment

PAH lifetime cancer risk estimates according to different authors are summarized in Table 4.19. Among the listed estimates, this study presents middle-level risks. Tuntawaroon et al. (2007) reported the highest risk of 1.7×10^{-4} for Bangkok schoolchildren based on personal exposure data. In Stockholm, the lifetime exposure of PAHs at a 3.0-ng/m^3-BaP equivalent concentration led to a lifetime cancer risk as high as 2.6×10^{-4} when the unit risk 8.7×10^{-5} was applied. In Sweden, PAH emission from domestic heating is dominant, and it is more than double the emission contribution by the transport sector. For Tokyo and Hokkaido, Japan, the risk estimates using the lower unit risk of 1.1×10^{-6} were on the order of 10^{-8} to 10^{-7}, which were acceptably low. Even when the unit risk of 8.7×10^{-5} was adopted, the risk estimated did not exceed the order of 10^{-5}. A comparison of the roadside ground level monitoring studies in Bangkok between this study and the one by Chetwittayachan (2002) finds that the former estimated the risk to be around 40 times higher, as shown in the table. However, if the same unit risk was applied, the latter would become double the value of the estimates in this study. This difference might be interpreted as the difference between sites in a city.

4.4 Conclusions

A risk assessment of urban air quality in street canyons was conducted in a case study on PAHs for two sites in Bangkok. Lifetime cancer risk was assessed for roadside residents by considering the diurnal and seasonal variations of PAH concentrations obtained from the field measurement. Lifetime risk was estimated to be significantly high at both sites, $5.6–6.7 \times 10^{-5}$ at the R6 site and $6.5–8.5 \times 10^{-5}$ at the CC site. Despite the clear difference in road environments, particularly the presence or absence of the elevated highway structure above the ground-level roads, a comparison of the two sites is not simple because of different meteorological and traffic conditions, the complex behaviors of PAHs, and sampling technical problems. Longer periods of monitoring data at the R6 site, where

Table 4.19 PAH cancer risk estimates in various case studies (Chetwittayachan 2002; Tuntawiroon et al. 2007; Kato et al. 2004; Burman 2001)

City	Bangkok	Bangkok	Bangkok	Tokyo	Hokkaido	Stockholm
PAH cancer risk estimate	$5.6–6.5 \times 10^{-5}$	1.7×10^{-4}	1.3×10^{-6}	1.0×10^{-6}	$0.99–3.7 \times 10^{-7}$	2.6×10^{-4}
Location	Roadside	Primary school and living area*	Roadside		Roof top of university building	City center
Year of monitoring	2005–2007	2004–2005	2001		2002	1997
Number of monitored PAHs	13 PAHs	10 PAHs	12 PAHs		5 PAHs	12 PAHs
BaP unit risk applied	8.7×10^{-5}	8.7×10^{-5}	1.1×10^{-6}		1.1×10^{-6}	8.7×10^{-5}
TEFs applied	Larsen and Larsen (1998)	Nisbet and Lagoy (1992)	California Environmental Protection (2002)		California Environmental Protection (2002)	Larsen and Larsen (1998)
Reference	This study	Tuntawiroon et al. (2007)	Chetwittayachan (2002)		Kato et al. (2004)	Burman (2001)

*Personal air samplers were used for 24 h for children going to primary schools

the wind was more stagnant in the street canyon due to the elevated structure of highway, would be expected to show higher pollution levels and higher health risk. The LLE was also calculated to show the PAH adverse effects in a dynamic way. The total LLEs were approximately 6.1 h at R6 and 7.1 h at CC. Through this case study, the methodology of airborne PAH risk assessment was also discussed, including its varieties and uncertainties, which should be better managed in the future for further improvement of urban air quality assessment.

References

Armstrong B, Tremblay C, Baris D, Thériault G (1994) Lung cancer mortality and polynuclear aromatic hydrocarbons: a case-cohort study of aluminium production workers in Arvida, Quebec, Canada. Am J Epidemiol 139(3):250–262

Asian Development Bank (2006) Country Synthesis Report on Urban Air Quality Management: Thailand. ADB, Manila

Baker JE, Eisenreich SJ (1990) Concentrations and fluxes of polycyclic aromatic hydrocarbons and polychlorinated biphenyls across the air-water interface of Lake Superior. Environ Sci Technol 24:342–352

Beak SO, Field RA, Goldstone ME, Kirk PW, Lester JN, Perry R (1991) A review of atmospheric polycyclic aromatic hydrocarbons: sources, fate and behavior. Water Air Soil Pollut 60:279–300

Boonyatumanond R, Murakami M, Wattayakorn G, Togo A, Takada H (2007) Sources of polycyclic aromatic hydrocarbons (PAHs) in street dust in a tropical Asian megacity, Bangkok, Thailand. Sci Tot Environ 384:420–432

Bostrom CE, Gerde P, Hanberg A, Jernstrom B, Johansson C, Kryklund T, Rannug A, Tornqvist M, Victorin K, Westerholm R (2002) Cancer risk assessment, indicators, and guidelines for polycyclic aromatic hydrocarbons in the ambient air. Environ Health Perspect 110:451–488

Brubaker WW, Hites RA (1998) OH reaction kinetics of polycyclic aromatic hydrocarbons and polychlorinated dibenzo-p-dioxins and dibenzofurans. J Phys Chem 102:915–921

Burman L (2001) The Air in Stockholm 2000: yearly report from SLB-analysis. Environmental and Health Protection Administration, Stockholm (in Swedish)

Burtscher H (1992) Measurement and characteristics of combustion aerosols with special consideration of photoelectric charging and charging by flame ions. J Aerosol Sci 23:549–595

Burtscher H, Schmidt-Ott A (1986) In situ measurement of adsorption and condensation of a polycyclic aromatic hydrocarbon on ultrafine c particles by means of photoemission. J Aerosol Sci 17:699–703

California Environmental Protection Agency (2002) Air toxics hot spots program risk assessment guidelines: Part II Technical support document for describing available cancer potency factors, California Environmental Protection Agency, California

CARB (1994) Benzo[a]pyrene as a toxic air concomitant; Part B Health assessment. Environmental Protection Agency, Air Resources Board, California

Chetwittayachan T (2002) Temporal variation of particle-bound polycyclic aromatic hydrocarbons (pPAHs) concentration and risk assessment of their possible human exposure in urban air environments. PhD thesis submitted to department of urban engineering, graduate school of engineering, The University of Tokyo, Tokyo

Chetwittayachan T, Shimazaki D, Yamamoto K (2002a) A comparison of temporal variation of particle-bound polycyclic aromatic hydrocarbons (pPAHs) concentration in different urban environments: Tokyo, Japan, and Bangkok, Thailand. Atmos Environ 36:2027–2037

Chetwittayachan T, Kido R, Shimazaki D, Yamamoto K (2002b) Diurnal profiles of particle-bound polycyclic aromatic hydrocarbon (pPAH) concentration in urban environment in Tokyo metropolitan area. Water Air Soil Pollut Focus 2:203–221

Collins JF, Brown SV, Marty MA (1991) Risk assessment for benzo[a]pyrene. Regul Toxicol Pharmacol 13:170–184

Ebert LB (1988) Polynuclear aromatic compounds. American Chemical Society, Washington, DC, pp 31–58

EcoChem Analytics (1999) User's guide: realtime PAH monitor PAS2000CE

Esteve W, Budzinski H, Villenave E (2006) Relative rate constants for the heterogeneous reactions of NO_2 and OH radicals with polycyclic aromatic hydrocarbons adsorbed on carbonaceous particles. Part 2: PAHs adsorbed on diesel particulate exhaust SRM 1650a. Atmos Environ 40:201–211

European Commission (2001) Ambient air pollution by Polycyclic Aromatic Hydrocarbons (PAH). Position Paper. Office for Official Publications of the European Communities, Luxembourg

European Environment Agency (2004) Air pollution in Europe 1990–2000. Topic report. European Environment Agency, Copenhagen

Feron VJ, DeJong D, Emmelot P (1973) Dose-response correlation for the induction of respiratory-tract tumours in Syrian golden hamsters by intratracheal instillations of benzo[a]pyrene. Eur J Cancer 9:387–390

Gamo M, Oka T, Nakanishi J (1996) Estimation of the loss of life expectancy from cancer risk due to exposure to carcinogens using life table. Environ Sci 9(1):1–8 [in Japanese]

Garivait H, Polprasert C, Yoshizumi K, Reutergardh LB (2001) Airborne polycyclic aromatic hydrocarbons (PAH) in Bangkok urban air: Part II. Level and distribution. Polycycl Aromat Comp 18:325–350

Heinrich U, Roller M, Pott F (1994) Estimation of a lifetime unit lung cancer risk for benzo[a]pyrene based on tumour rates in rats exposed to coal tar/pitch condensation aerosol. Toxicol Lett 72:155–161

Hoshiko T, Yamamoto K, Fukushi K, Chetwittayachan T (2005) Temporal and spatial variation of particle-bound polycyclic aromatic hydrocarbons (PAHs) concentration and PAH compositions at different particle-size ranges in urban air environment in Bangkok metropolitan area. Air Pollution XIII: 551–562

IARC (2011) IARC monographs on the evaluation of carcinogenic risks to humans. IARC, Lyon. http://monographs.iarc.fr/ENG/Classification/index.php. Accessed Jul 27, 2011

IPCS (1998) Selected non-heterocyclic polycyclic aromatic hydrocarbons. Environmental Health Criteria 202. WHO, Geneva

Jiang T, Oooka R, Kato S, Hwang K, Takahashi K, Kouno R, Watanabe S (2004) Study on prediction of pollutant dispersion in urban area (Part 3): numerical prediction of pollutants dispersion by traffic road. In: Proceeding of the conference of Japan Architecture Society, in Hokkaido, August 2004 [in Japanese]

Kameda Y, Shirai J, Komai T, Nakanishi J, Masunaga S (2005) Atmospheric polycyclic aromatic hydrocarbons: size distribution, estimation of their risk and their depositions to the human respiratory tract. Sci Tot Environ 340:71–80

Kamens RM, Guo Z, Fulcher JN, Bell DA (1988) Influence of humidity, sunlight, and temperature on the daytime decay of polycyclic aromatic hydrocarbons on atmospheric soot particles. Environ Sci Technol 22:103–108

Kato H, Murao N, Yamagata S, Ohta S (2004) Estimation of cancer risk from diesel exhaust particle (DEP) based on NOx data at air monitoring stations in Sapporo, Japan. Environ Eng Res 41:111–117 [in Japanese]

Kim DY, Yamaguchi K, Kondo A, Soda S (2001) Study on relationship between photochemical oxidant concentration and primary pollutants emission amounts in Osaka and Hyogo regions. J Jpn Soc Atmos Environ 36:156–165 [in Japanese]

Larsen JC, Larsen PB (1998) Chemical carcinogens. In: Hester RE, Harrison RM (eds) Air pollution and health. The Royal Society of Chemistry, Cambridge

Lohmann R, Northcott GL, Jones KC (2000) Assessing the contribution of diffuse domestic burning as a source of PCDD/Fs, PCBs, and PAHs to the UK atmosphere. Environ Sci Technol 34:2892–2899

Marple VA, Rubow KL, Behm SM (1991) A microorifice uniform deposit impactor (MOUDI): description, calibration and use. Aerosol Sci Technol 14:434–446

MSP Corporation (1998) Micro-orifice uniform deposit impactor instruction manual, Model 100/ Model 110. MSP Corporation, Shoreview

Muller P (1997) Scientific criteria document for multimedia standards development polycyclic aromatic hydrocarbons (PAH); Part 1: Hazard identification and dose-response assessment. Standard Development Branch, Ontario Ministry of Environment and Energy, Ontario

Nielsen T (1996) Traffic contribution of polycyclic aromatic hydrocarbons in the center of a large city. Atmos Environ 30:3481–3490

Nielsen T, Jorgensen HE, Larsen JC, Poulen M (1996) City air pollution of polycyclic aromatic hydrocarbons and other mutagens: occurrence, sources and health effects. Sci Tot Environ 189 (190):41–49

Nisbet IT, Lagoy PK (1992) Toxic equivalency factors (TEFs) for polycyclic aromatic hydrocarbons (PAHs). Regul Toxicol Pharm 16:290–300

Pankow JF, Bidleman TF (1991) Effects of temperature, TSP and per cent non-exchangeable material in determining the gas-particle partitioning of organic compounds. Atmos Environ 25:2241–2249

Panther B, Hooper M, Limpaseni W, Hooper B (1996) Polycyclic aromatic hydrocarbon as environmental contaminants: some results from Bangkok. The third international symposium of ETERNET-APR: conservation of the hydrospheric environment, Bangkok. December 1996

Panther B, Hooper MA, Tapper NJ (1999) A comparison of air particulate matter and associated polycyclic aromatic hydrocarbons in some tropical and temperate urban environments. Atmos Environ 33:4087–4099

Pierce RC, Katz M (1975) Dependency of polynuclear aromatic hydrocarbon content on size distribution of atmospheric aerosols. Environ Sci Technol 9:343–353

Pike MC (1983) Human-cancer risk assessment. In: Commission on Life Sciences (ed) Polycyclic aromatic hydrocarbons: evaluation of sources and effects. National Academy Press, Washington, DC

Pollution Control Department, Thailand (2004) Thailand state of pollution report 2004. Pollution Control Department, Bangkok

Pope CA III, Dockery DW (1999) Epidemiology of particle effects. In: Holgate ST, Koren HS, Samet JM, Maynard RL (eds) Air pollution and health. Academic, San Diego

Pott F (1985) Pyrolyseabgase PAH, Lungkrebsrisiko - daten und bewertung. STAUB-Reinhalt Luft 45(7/8):369–379 [in German]

Prevedouros K, Jones KC, Sweetman AJ (2004) Modelling the atmospheric fate and seasonality of polycyclic aromatic hydrocarbons in the UK. Chemosphere 56:195–208

Raat WK, Bakker GL, Meijere FA (1990) Comparison of filter materials used for sampling of mutagens and polycyclic aromatic hydrocarbons in ambient airborne particles Original Research Article. Atmos Environ 24:2875–2887

RIVM (1989) Integrated criteria document PAHs. Bilthoven: National Institute of Public Health and Environmental Protection. 758474011: pp 1–199

Ruchirawat M, Mahidol C, Tangjarukij C, Pui-ock S, Jensen O, Kampeerawipakorn O, Tuntaviroon J, Aramphongphan A, Autrup H (2002) Exposure to genotoxins present in ambient air in Bangkok, Thailand—particle associated polycylic aromatic hydrocarbons and biomarkers. Sci Tot Environ 287:121–132

Saffiotti U, Montesano R, Sellakumar AR, Kaufman DG (1972) Respiratory tract carcinogenesis induced in hamsters by different dose levels of benzo[a]pyrene and ferric oxide. J Natl Cancer Inst 49:1199–1204

Tamamura S, Sato T, Ota Y, Wang X, Tang N, Hayakawa K (2007) Long-range transport of polycyclic aromatic hydrocarbons (PAHs) from the eastern Asian continent to Kanazawa, Japan with Asian dust. Atmos Environ 41:2580–2593

Thyssen J, Althoff J, Kimmerle G, Mohr U (1981) Inhalation studies with benzo[a]pyrene in Syrian golden hamsters. J Natl Cancer Inst 66(3):575–577

Tuntawiroon J, Mahidol C, Navasumrit P, Autrup H, Ruchirawat M (2007) Increased health risk in Bangkok children exposed to polycyclic aromatic hydrocarbons from traffic-related sources. Carcinogenesis 28(4):816–822

Venkataraman C, Friedlander SK (1994) Size distribution of polycyclic aromatic hydrocarbons and elemental carbon. 2. Ambient measurements and effects of atmospheric processes. Environ Sci Technol 28:563–572

Venkataraman C, Lyons JM, Friendlander SK (1994) Size distributions of polycyclic aromatic hydrocarbons and elemental carbon. 1. Sampling, measurement methods, and source characterization. Environ Sci Technol 28:555–562

Venkataraman C, Thomas S, Kulkarni P (1999) Size distributions of polycylic aromatic hydrocarbons—gas/particle partitioning to urban aerosols. J Aerosol Sci 30:759–770

Westerholm R, Stenberg U, Alsberg T (1988) Some aspects of the distribution of polycyclic aromatic hydrocarbons (pah) between particles and gas phase from diluted gasoline exhausts generated with the use of a dilution tunnel, and its validity for measurement in ambient air. Atmos Environ 22:1005–1010

WHO (1987) Air quality guidelines for Europe, WHO Regional Publications, European series No. 23. WHO, Copenhagen

WHO (2000) Air quality guidelines for Europe, 2nd edn, WHO Regional Publications, European series No. 91. WHO, Copenhagen

WHO (2002) World health report 2002; Reducing risks, promoting life. WHO, Geneva. http://www.who.int/whr/2002/en/index.html. Accessed May 5, 2008

Wiest F, Fiorentina HD (1975) Suggestions for a realistic definition of an air quality index relative to hydrocareonaceous matter associated with airborne particles. Atmos Environ 9:951–954

Chapter 5
Pollutant Dispersion in an Urban Area

Keisuke Nakao and Shinsuke Kato

Abstract This chapter is composed by an introduction of existing prediction methodology of pollutant transportation and experimental observations of concentration fluctuation. In Sect. 5.2, the convective diffusion equation is derived from a balance of mass transportation, and approximated expression of average concentration theoretically derived from the transport equation is introduced. In Sect. 5.3, the results of wind tunnel experiment on downscaled model of existent urban area are indicated. The several characteristic points in urban area, such as highway, school ground, and in-between space of high-rise buildings, are selected as an occurrence point of pollutant. Instantaneous concentration is measured in-between of office buildings. The time of transport between source and measuring point is measured. The property of instantaneous concentration is discussed by focusing the probability density function (PDF) and the higher order moments, skewness, and kurtosis. These properties are linked to the distance between source and measuring point and condition of buildup area.

Keywords Urban air pollution • Unsteady emission • Higher order moment • Probability density function • Gaussian plume model

K. Nakao (✉)
Faculty of Engineering, The University of Tokyo, 4-6-1,
Komaba, Meguro-ku, Tokyo 1538505, Japan
e-mail: knakao@iis.u-tokyo.ac.jp

S. Kato
Institute of Industrial Science, The University of Tokyo, 4-6-1,
Komaba, Meguro-ku, Tokyo 1538505, Japan
e-mail: kato@iis.u-tokyo.ac.jp

S. Kato and K. Hiyama (eds.), *Ventilating Cities: Air-flow Criteria*
for Healthy and Comfortable Urban Living, Springer Geography,
DOI 10.1007/978-94-007-2771-7_5, © Springer Science+Business Media B.V. 2012

5.1   Introduction

This chapter summarizes the diffusion phenomenon in the urban area. The pollution in the urban area caused by material adverse to human health is a problem that appeared along with industrial development. For example, continuous pollutant emission from chimneys results in prolonged exposure for the local residents. It could also cause harm to the living environment around the facility. The primary countermeasure to these problems is source control. Source control is a method that prevents emission from the generation source. It is in fact a countermeasure that requires performance improvement in industrial technology rather than in air environment control. It is essential in terms of eradicating the harmful cause. In addition, much knowledge on dilution and diffusion of material has been obtained through experiments and measurements. In many cases, an approximate model is used to predict material diffusion in the urban area. These models are based on a theoretical formula of diffusion. Therefore, the advection diffusion equation of general material is derived in Sect. 5.2. That section also introduces the summary of an approximate model based on the equation and its derivation. This approximate solution is highly versatile because it considers a variety of pollutant generation conditions and assumes various formats. Therefore, it is still often used for the diffusion phenomenon of chimney exhaust from factories. However, the solution is not necessarily formed only by a theoretical formula. In the phase of developing the prediction models, some part of the properties of turbulent flow were abandoned, although the effect of which is not negligible. Numerous researchers have performed measurements and experiments and developed the approximated expression and adjusted variables so that the approximation matches the actual phenomenon. Thus, approximation is accurate only when the actual phenomenon is similar to the experimental conditions used.

Here, two points must be considered: that this approximation is valid when the ground surface is a flat surface or nearly so and that the concentration response is averaged.

In contrast to the conventional air pollution due to factory exhausts, recent air pollution problems arise due to exhaust gas generated in city areas or the sudden occurrence of toxic pollutants. The two factors stated above present huge detriments in applying these models to recent pollution problems.

Currently, diffusion phenomena must assume nonstationary material generation and realistic reflection of the influence of city buildings as obstacles when considering the leakage of biological weapons or hazardous chemicals by terrorism or accident in cities. The main target of discussion for the approximation mentioned above is the average exposure to a toxic substance. Regarding the exhaust from chimneys, the chronic accumulation of toxic substances in the human body was thought to be mitigated by metabolism if the chimney height was sufficient and the city area was proven to be a low concentration region by approximated prediction model. Thus, the conventional assessment method of the diffusion phenomenon was to determine an appropriate chimney height based on the approximation model. However, the current problem arises due to a high concentration generated suddenly near the ground surface and not due to the low concentration from chimneys over the long term. In the case of the leakage of a toxic substance due to terrorism or an accident, an emergency situation, such as inhaling the high concentration toxic substance, could exceed human tolerance limits in a short time, which cannot be described by the average concentration mentioned above. Thus, the time history of an instantaneous concentration must be investigated. However, the transport equation of flow includes nonlinear properties in the mechanisms of air transport. These properties vary the transport of concentration from moment to moment. Especially in cities with numerous buildings, in contrast to the flat ground surface, the scale of vortices transporting the concentration differs significantly depending on the location. Therefore, the transport of concentration becomes highly complicated and defies a simple prediction. In this case, the fact that the risk differs depending on time and location makes the application to these emergency situations even more difficult.

In Sect. 5.3, the actual city shape is simulated. Discussion regarding the wind tunnel test using tracer gas assuming the sudden occurrence of toxic substance due to terrorism or an accident is included in this section. The instantaneous concentration response is recorded by locating the measurement points in between buildings, which is small open space in urban areas, and emitting tracer gas from each generation point to represent the main characteristics of existing city blocks.

5.2 Approximate Model of Material Diffusions Phenomena in the Atmosphere

This section describes the equation for simplified concentration prediction. First, the molecular scale transport by Fick's law is introduced. Then, the concentration transport equation of the material transported in the fluid is derived. The averaging procedure is performed for a simplified prediction of concentration, and, in the end, the hypothesis to perform the approximation of the term due to vortices is presented.

5.2.1 Fick's Law

The substances are assumed to be mixed in the volume. The amount of substance has dimensions of [M] (length is [L], and time is [T]). When the substance is present, the ratio of the substance in the volume can be defined, which is termed concentration. The concentration has dimensions of $[ML^{-3}]$ as amount per volume.

The law applies to the transport amount of substance per unit time (dimensions are $[ML^{-2} T^{-1}]$) that passes through the arbitrarily located unit surface. When regions with high concentration and low concentration exist in the same space, the substance is transported from the high concentration region to the low concentration region by molecular diffusion. The transport amount is determined by the concentration gradient, dC/dx_i, which is Fick's law. The concentration is expressed as $C[ML^{-3}]$. Figure 5.1 illustrates the conceptual scheme.

Fig. 5.1 Transport by
concentration gradient

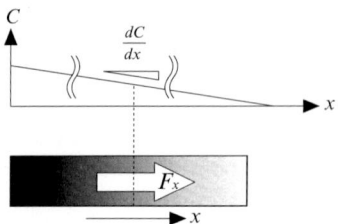

According to Fick's law, the transport amount of substance per unit time that passes through the unit surface $F[\mathrm{ML^{-2}\,T^{-1}}]$ is expressed as follows:

$$F = D\left(-\frac{dC}{dx}\right) \qquad (5.1)$$

Here, D is diffusion coefficient, and its dimensions are $[\mathrm{L^2T^{-1}}]$. This value differs by substance.

5.2.2   Application of Fick's Law for Diffusion in Fluids

Here, Fick's law is applied in the flow field. There are two ways to observe the fluid phenomenon. One is to fix the observation space and consider the balance of physical quantities with respect to the fluid that passes through this space. This method is based on the Euler perspective. The other method is to view the scale of fluid vortex as one lump (fluid body) and to capture it as the motion of a body. This method is based on the Lagrange perspective. For example, when many balls are released in the river, observing the number of balls that passes through one location is the Euler perspective, and investigating the travel distance and/or velocity of one ball is the Lagrange perspective. In this section, the movement of one fluid body in the flow is considered based on the Lagrange perspective.

The image of fluid body movement is shown in Fig. 5.2. When the fluid body occupying the space $(\delta x_1, \delta x_2, \delta x_3)$ (dimension [L]) is located at (x_1, x_2, x_3) at time t (dimension [T]), it has a concentration of $C(x_1, x_2, x_3, t)$. After timeΔt, the previous fluid body is displaced by $(\Delta x_1, \Delta x_2, \Delta x_3)$, and its concentration is now given by$C(x_1 + \Delta x_1, x_2 + \Delta x_2, x_3 + \Delta x_3, t + \Delta t)$. If the concentration is a field function controlled by x_1, x_2, x_3, t, the capacity of variation in concentration that corresponds to the change in field $(\Delta x_1, \Delta x_2, \Delta x_3, \Delta t)$ is expressed approximately by the summation of the partial differentiation of each physical quantity and the product of small variation capacity.

$$\Delta C = \frac{\partial C}{\partial t}\cdot \Delta t + \frac{\partial C}{\partial x_1}\cdot \Delta x_1 + \frac{\partial C}{\partial x_2}\cdot \Delta x_2 + \frac{\partial C}{\partial x_3}\cdot \Delta x_3 \qquad (5.2)$$

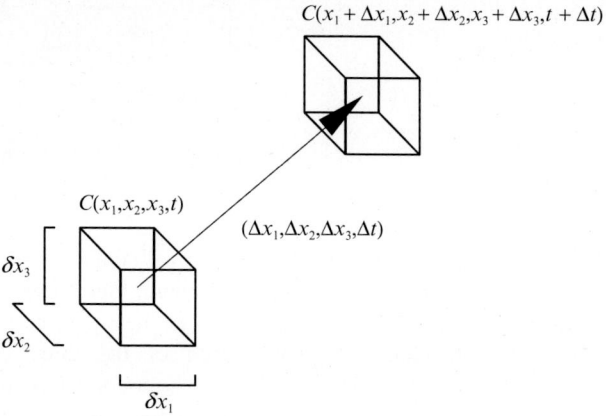

Fig. 5.2 Movement of the fluid body

In fact, this equation yields dimensions of concentration. To understand the changes in the amount of substance, the volume must be integrated. However, because the change in fluid volume and its density during the movement is not considered here, the amount of change in substance can be examined by the amount of change in concentration.

To express this as rate of change per unit time, both sides are divided by Δt, and the time interval is reduced to the utmost limit. It is expressed as

$$
\lim_{\Delta t \to 0} \frac{\Delta C}{\Delta t} = \lim_{\Delta t \to 0} \left(\frac{\partial C}{\partial t} + \frac{\partial C}{\partial x_1} \cdot \frac{\Delta x_1}{\Delta t} + \frac{\partial C}{\partial x_2} \cdot \frac{\Delta x_2}{\Delta t} + \frac{\partial C}{\partial x_3} \cdot \frac{\Delta x_3}{\Delta t} \right)
$$
$$
= \frac{\partial C}{\partial t} + u_1 \frac{\partial C}{\partial x_1} + u_2 \frac{\partial C}{\partial x_2} + u_3 \frac{\partial C}{\partial x_3} \tag{5.3}
$$

where (u_1, u_2, u_3) is the velocity (dimension $[LT^{-1}]$). This form is called the substantial derivative and is formally denoted as

$$
\frac{DC}{Dt} \tag{5.4}
$$

When the fluid body is changed temporally and spatially by $(\Delta x_1, \Delta x_2, \Delta x_3, \Delta t)$, the amount of change of substance in the fluid body is expressed as

$$
\frac{DC}{Dt} \cdot \delta x_1 \cdot \delta x_2 \cdot \delta x_3 \cdot \Delta t \tag{5.5}
$$

Here, the change in the amount of substance was induced by the concentration gradient transport according to Fick's law. By assuming the transport of an amount of substance in unit time of the volume surface, determined as the fluid body,

Fig. 5.3 Balance of
substance transport in the test
volume

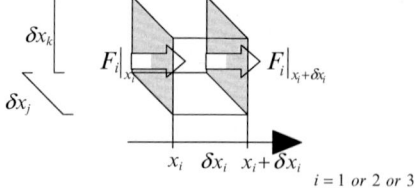

and each unit surface as F_i, this amount is expressed with tensor notation using the
concentration gradient and diffusion coefficient D_{ij} (here, Fick's law is extended to
be expressed as a contraction of three components). In the equation, i indicates the
direction of transportation and j indicates each of the three directions of
components that contribute to transportation in the i direction. Here, j is a contraction. Contraction means that when the letter appears in the equation with subscripts,
such as (i, j, k) (called suffix), and when it appears multiple times in one term, the
summation is computed over all possible values of i:

$$F_i = D_{ij}\left(-\frac{\partial C}{\partial x_j}\right) = D_{i1}\left(-\frac{\partial C}{\partial x_1}\right) + D_{i2}\left(-\frac{\partial C}{\partial x_2}\right) + D_{i3}\left(-\frac{\partial C}{\partial x_3}\right) \qquad (5.6)$$

The equation will be derived logically from the balance between the change of
concentration in the fluid body occupying test volume $(\delta x_1 \cdot \delta x_2 \cdot \delta x_3)$ and the
amount of substance flowing in to and out from the surface of the fluid body;

$$-\Delta t \cdot \left\{\left(F_1|_{x_1+\delta x_1} - F_1|_{x_1}\right)\delta x_2 \cdot \delta x_3 + \left(F_2|_{x_2+\delta x_2} - F_2|_{x_2}\right)\delta x_1 \cdot \delta x_3\right.$$
$$\left. + \left(F_3|_{x_3+\delta x_3} - F_3|_{x_3}\right)\delta x_1 \cdot \delta x_2\right\} \qquad (5.7)$$

The expression calculates the sum of substances flowing in to and out of each
surface. The image is shown in Fig. 5.3. Here, for simplified understanding, only the
first term of Eq. 5.7 is considered:

$$-\Delta t \cdot \left(F_1|_{x_1+\delta x_1} - F_1|_{x_1}\right)\delta x_2 \cdot \delta x_3 \qquad (5.8)$$

Equation 5.8 becomes

$$-\Delta t \cdot \left(F_1|_{x_1+\delta x_1} - F_1|_{x_1}\right)\delta x_2 \cdot \delta x_3$$
$$= -\Delta t \cdot \left\{\left\{D_{1j}\left(-\frac{\partial C}{\partial x_j}\right)\right\}\bigg|_{x_1+\delta x_1} - \left\{D_{1j}\left(-\frac{\partial C}{\partial x_j}\right)\right\}\bigg|_{x_1}\right\}\delta x_2 \cdot \delta x_3$$
$$= -\Delta t \cdot \frac{\left\{D_{1j}\left(-\frac{\partial C}{\partial x_j}\right)\right\}\bigg|_{x_1+\delta x_1} - \left\{D_{1j}\left(-\frac{\partial C}{\partial x_j}\right)\right\}\bigg|_{x_1}}{\delta x_1}\delta x_1 \cdot \delta x_2 \cdot \delta x_3 \qquad (5.9)$$

Here, j is a contraction. The case for $i = 1$ is shown here. However, the equation can be developed in the same manner for the other cases. Equation 5.7 becomes

$$- \Delta t \cdot \sum_{i=1,2,3} \frac{\left\{ \left\{ D_{ij}\left(-\frac{\partial C}{\partial x_j}\right)\right\}\big|_{x_i+\delta x_i} - \left\{ D_{ij}\left(-\frac{\partial C}{\partial x_j}\right)\right\}\big|_{x_i} \right\}}{\delta x_i} \delta x_1 \cdot \delta x_2 \cdot \delta x_3 \qquad (5.10)$$

Here, i and j are both contractions. Equation 5.7 is assumed to be equal to Eq. 5.5:

$$\frac{DC}{Dt} \cdot \delta x_1 \cdot \delta x_2 \cdot \delta x_3 \cdot \Delta t$$

$$= -\Delta t \cdot \sum_{i=1,2,3} \frac{\left\{ \left\{ D_{ij}\left(-\frac{\partial C}{\partial x_j}\right)\right\}\big|_{x_i+\delta x_i} - \left\{ D_{ij}\left(-\frac{\partial C}{\partial x_j}\right)\right\}\big|_{x_i} \right\}}{\delta x_i} \delta x_1 \cdot \delta x_2 \cdot \delta x_3$$

$$\qquad\qquad (5.11)$$

The space dimensions obtained by the fluid body, $\delta x_1, \delta x_2, \delta x_3$, are reduced as far as possible:

$$\frac{DC}{Dt} = \frac{\partial}{\partial x_i}\left\{ D_{ij}\left(\frac{\partial C}{\partial x_j}\right)\right\} \qquad (5.12)$$

The advection diffusion equation is now derived from the above equations. Intrinsically, the coefficient of molecular diffusion is included in partial differentiation. However, when considering the diffusion of the substance in the fluid, the diffusion coefficient tensor is assumed such that only diagonal components have a nonzero value. The coefficient of molecular diffusion D_m is used as the value of diffusion coefficient, assuming that it is constant, which does not cause a large discrepancy in the phenomenon. The expression is simplified to

$$\frac{\partial C}{\partial t} + u_i \frac{\partial C}{\partial x_i} = D_m \frac{\partial^2 C}{\partial x_i^2} \qquad (5.13)$$

This equation expresses the transport of pollutant gas in the atmosphere. Here, i is a contraction. The first term on the left side of this equation indicates a temporal progress of concentration. The second term on the left side indicates the transport of concentration by advection. The first term on the right side indicates molecular diffusion. This form was derived based on the Lagrange perspective. However, the final results from the Euler's perspective will be the same.

5.2.3 *Application of Substance Diffusion Phenomena in the Atmosphere*

This subsection summarizes the approximate model of substance diffusion in an urban area that is used to predict urban area diffusion.

First, the atmosphere is generally considered to be in a condition of turbulence. Under this condition, the concentration fluctuates greatly due to the nonlinearity of the flow transport equation and complexity of concentration transport.

The long-term exposure from chimneys was a main subject of recent air pollution issues. As a countermeasure, it is very important to reduce the amount of pollutants that are generated. This does not imply that the air environment is to be evaluated. It is a solution that focuses on investigating the improvement of efficiency in industrial technology. The atmosphere is only important in terms of concentration properties when the diffusion from chimneys reaches the ground surface. Therefore, based on the transport equation of concentration mentioned above, a simplified method to calculate the concentration in a downstream basin and on the ground surface was determined.

However, the concentration transport under turbulent conditions is extremely complex. Therefore, the equation must be simplified for prediction.

The danger of long-term exposure to chimney exhaust on human health is usually determined as an accumulated amount. In other words, the problem focuses on how much pollutant remains in the human body, which is not metabolized, when exposed to a certain average concentration environment over the long term. Thus, the deviation in the concentration fluctuation is not a serious problem. The health hazard due to being exposed to an average concentration for years is the main concern. The following discussion is based on the assumption that an averaging procedure has been performed.

The average observed concentration value is denoted as $\bar{C}$, and the gap between the observation value and the average is denoted as C'. In addition, the average velocity is denoted as $\bar{u}_i$, and the gap from the average is denoted as u'_i, which is called the Reynolds decomposition. These terms are expressed as

$$C = \bar{C} + C'$$
$$u_i = \bar{u}_i + u'_i \tag{5.14}$$

However,

$$\overline{C'} = 0$$
$$\overline{u'_i} = 0 \tag{5.15}$$

The Reynolds decomposition is considered one kind of decomposition using average quantity by ensemble average. The ensemble average is determined by averaging the results of repeated trials under the same condition. For example, assume that a valve is filled with pressurized gas. The gas is emitted by opening the valve, and the physical quantity of the released gas is measured at a fixed location and time. If the trial is repeated under the same pressure condition and opening condition, the same number of data for the gas at same location and time is obtained as the number of trials. The results of these trials are slightly different due to the small differences in the initial conditions. However, the measurement results are utilized to approach one value by increasing the number of trials and performing an averaging procedure. For example, consider the concentration C. Assuming that the measurement was performed n times, the average result and the gap from it are expressed as

$$\bar{C} = \frac{1}{n} \sum_n C_k$$

$$C_k' = C_k - \frac{1}{n} \sum_n C_k \tag{5.16}$$

which shows that Eqs. 5.14 and 5.15 are true. The Reynolds decomposition and Reynolds average imply that the values would converge to one value if the trial was repeated infinitely.

Each term that has contributed to the Reynolds decomposition based on the above is substituted into the advection diffusion equation. Using the assumption of Reynolds decomposition, each term is incorporated as follows:

$$\frac{\overline{\partial C}}{\partial t} = \frac{\overline{\partial (\bar{C} + C')}}{\partial t} = \frac{\partial \bar{C}}{\partial t} \tag{5.17}$$

$$\overline{u_i \frac{\partial C}{\partial x_i}} = \overline{(\bar{u}_i + u_i') \frac{\partial (\bar{C} + C')}{\partial x_i}}$$

$$= \bar{u}_i \frac{\partial \bar{C}}{\partial x_i} + \bar{u}_i \overline{\frac{\partial C'}{\partial x_i}} + \overline{u_i' \frac{\partial \bar{C}}{\partial x_i}} + \overline{u_i' \frac{\partial C'}{\partial x_i}} = \bar{u}_i \frac{\partial \bar{C}}{\partial x_i} + \overline{u_i' \frac{\partial C'}{\partial x_i}} \tag{5.18}$$

Incidentally, using the continuous equation, the right side is expressed as

$$\overline{u_i' \frac{\partial C'}{\partial x_i}} = \frac{\overline{\partial u_i' C'}}{\partial x_i} - \overline{C' \frac{\partial u_i'}{\partial x_i}} = \frac{\overline{\partial u_i' C'}}{\partial x_i} \tag{5.19}$$

Moving this term from the right side to the left side, Eq. 5.13 becomes

$$\frac{\partial \bar{C}}{\partial t} + \bar{u}_1 \frac{\partial \bar{C}}{\partial x_1} + \bar{u}_2 \frac{\partial \bar{C}}{\partial x_2} + \bar{u}_3 \frac{\partial \bar{C}}{\partial x_3}$$

$$= D_m \frac{\partial^2 \bar{C}}{\partial x_1^2} + D_m \frac{\partial^2 \bar{C}}{\partial x_2^2} + D_m \frac{\partial^2 \bar{C}}{\partial x_3^2} + \frac{\partial \left(-\overline{u_1' C'}\right)}{\partial x_1} + \frac{\partial \left(-\overline{u_2' C'}\right)}{\partial x_2} + \frac{\partial \left(-\overline{u_3' C'}\right)}{\partial x_3} \tag{5.20}$$

The first term on the left side characterizes the temporal progress of concentration. The second, third, and fourth terms on the left side indicate the average concentration transport by the averaged flow field. The first, second, and third terms on the right side represent the molecular diffusion. The fourth, fifth, and sixth terms on the right side are the averages of the correlation terms of variation. The last four terms appear along with the Reynolds average of advection terms. The term $\overline{u'_i C'}$ is an unknown variable, as are $\bar{u}_i$ and $\bar{C}$. Performing the Reynolds average increases the number of variables and complexity of the equation. Thus, to eliminate the complexity, modeling the term with a manageable number of variables was attempted to reduce the number of variables. These correlation terms promote the mixture of substances; therefore, they are thought to have a diffusion effect. As mentioned in Fick's law, the diffusion effect indicates an effect where the substances move to reduce the inclination in the space that leads to a high-concentration region and a low-concentration region. A simplified modeling of this term based on the gradient transport of concentration was attempted. The following shows the modeling of the diffusion effect due to turbulence:

$$\overline{u'_1 C'} = -K_1 \frac{\partial \bar{C}}{\partial x_1}, \quad \overline{u'_2 C'} = -K_2 \frac{\partial \bar{C}}{\partial x_2}, \quad \overline{u'_3 C'} = -K_3 \frac{\partial \bar{C}}{\partial x_3} \qquad (5.21)$$

The coefficient K_i is called the turbulence diffusion coefficient, which is based on the analogy with Fick's law. However, it is only an analogy, and such gradient transport is not the principal property that the fluctuation correlation term displays in reality. The origin of the fluctuation correlation term is in the advection term, or in other words, a term that appears along with the movement of the fluid body. The derivation process of the advection diffusion equation shows that the absence of gradient transport by molecular viscosity (when Eq. 5.7, which becomes the right side of Eq. 5.13, does not exist) indicates that there is no concentration change with fluid body movement (Eq. 5.5, which becomes the left side of the equation). In other words, if there is no molecular diffusion, the fluid body with the concentration C_0 as an initial condition does not change in concentration with time, and it will never obtain a change in concentration. In this point, the effect of the turbulent diffusion and that of molecular diffusion is quite different. The turbulent diffusion effect means that the test volume assumes a large displacement, and therefore, the test volume with concentration C_0 is observed over the whole space. In that phase, the concentration gradient that was partial in the initial phase is observed over the whole space, and it becomes a trigger for realizing a space with a variety of concentrations with molecular diffusion over a large region. Generally, its effect of mixing is extremely high compared to molecular diffusion in the field with developed turbulence. Thus, though it does not have a diffusion effect in a strict sense, it can still promote diffusion. Thus, the assumption such as that used in Eq. 5.19 was made. Here, the important point is that for partial differentiation, the variable C, which was an instantaneous concentration in Fick's equation, is substituted with $\bar{C}$ in the approximation of the turbulent diffusion term. When the turbulence of the field is sufficiently developed and the turbulent diffusion effect contributes greatly to the diffusion phenomenon, ignoring the molecular diffusion term because it is small compared to turbulent diffusion, the equation becomes

$$\frac{\partial \bar{C}}{\partial t} + \bar{u}_1 \frac{\partial \bar{C}}{\partial x_1} + \bar{u}_2 \frac{\partial \bar{C}}{\partial x_2} + \bar{u}_3 \frac{\partial \bar{C}}{\partial x_3} = K_1 \frac{\partial^2 \bar{C}}{\partial x_1^2} + K_2 \frac{\partial^2 \bar{C}}{\partial x_2^2} + K_3 \frac{\partial^2 \bar{C}}{\partial x_3^2} \qquad (5.22)$$

The equation is even simpler here because the fluctuation correlation term, which was unknown before, has been substituted by the slope of the average concentration. However, each turbulent diffusion coefficient K_i is not constant because the scales of various vortices of turbulence have an influence on the concentration transport. The values are uncertain because they vary depending on the scale of vortex of turbulence. The examination will continue with the assumption that K_i is constant in terms of time and location. Eliminating the inconsistency created by this assumption motivates the approximation of the diffusion.

Here, x_1, x_2, x_3 will each be described as corresponding coordinates, x, y, z. If the mean flow does not exist ($\bar{u}_i = 0$), the left side of (5.22) takes on a simplified form. The primary diffusion equation will also be considered as a simplified one-dimensional model for understanding the actual three-dimensional phenomenon:

$$\frac{\partial \bar{C}}{\partial t} = K_x \frac{\partial^2 \bar{C}}{\partial x^2} \qquad (5.23)$$

The solution to this equation can be obtained by separating variables or by a Fourier transform with proper boundary conditions.

Let us consider the situation that the substance instantaneously incurred an increment of Q (dimension [M]) from $x = 0$.

The solution is

$$\bar{C} = \frac{Q}{\sqrt{4\pi Kt}} \exp\left(-\frac{x^2}{4K_1 t}\right) \qquad (5.24)$$

Equation 5.25 is the equation of Gaussian distribution of the statistics. Here, μ is the average value, and σ is the standard deviation:

$$f(x) = \frac{1}{\sqrt{2\pi}\sigma} \exp\left(-\frac{(x-\mu)^2}{2\sigma^2}\right) \qquad (5.25)$$

The distribution form of σ with $\mu = 0$ is shown in Fig. 5.4. The figure shows the result of changing σ from 0.05 to 2. It can be seen that the expansion of the distribution profile becomes larger as σ increases. In a Gaussian distribution, approximately 95% of the surface in the x region of $\pm 2\sigma$ with a central focus on the average value (here, it is 0) is occupied. Thus, 95% of all possible values are included in this range for the probability density distribution.

Fig. 5.4 Gaussian
distribution function

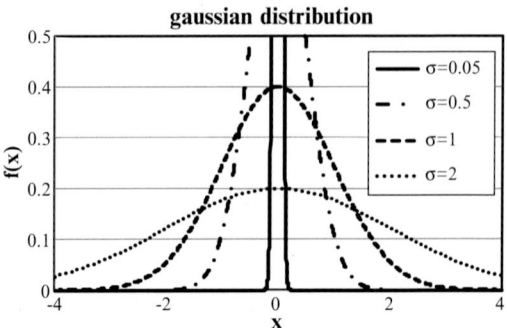

Based on this result, (5.22) is the Gaussian distribution corresponding to $\sigma = \sqrt{2K_x t}$, when excluding Q from the examination target. Here, t indicates the elapsed time. In the other words, the expansion range increases along with time.

Consider the case where the uniform velocity of U exists as the main stream only in the x direction:

$$\frac{\partial \bar{C}}{\partial t} + U \frac{\partial \bar{C}}{\partial x} = K_1 \frac{\partial^2 \bar{C}}{\partial x^2} \tag{5.26}$$

Assume instantaneous pollution occurrence conditions. The conversions $X = x - Ut$, $T = t$ are performed (Galilean transformation):

$$\frac{\partial \bar{C}}{\partial t} = \frac{\partial \bar{C}}{\partial T} \frac{\partial T}{\partial t} + \frac{\partial \bar{C}}{\partial X} \frac{\partial X}{\partial t} = \frac{\partial \bar{C}}{\partial T} - U \frac{\partial \bar{C}}{\partial X} \tag{5.27}$$

$$U \frac{\partial \bar{C}}{\partial x} = U \left(\frac{\partial \bar{C}}{\partial X} \frac{\partial X}{\partial x} + \frac{\partial \bar{C}}{\partial T} \frac{\partial T}{\partial x} \right) = U \frac{\partial \bar{C}}{\partial X} \tag{5.28}$$

$$K_1 \frac{\partial^2 \bar{C}}{\partial x_1^2} = K_1 \frac{\partial^2 \bar{C}}{\partial X^2} \tag{5.29}$$

Combining these equations, (5.22) is expressed in the form where x is substituted by X, and the solution can be obtained as follows:

$$\bar{C} = \frac{Q}{\sqrt{4\pi K_1 t}} \exp \left(-\frac{(x - Ut)^2}{4K_1 t} \right) \tag{5.30}$$

The barycentric position of concentration, which was 0 in the previous equation, is transiting in the U direction under the influence of advection in the x direction, and the substance is expanding. (Imagine the distribution in Fig. 5.4 is transiting toward the U direction as σ increases along with time.)

Now, the three-dimensional advection diffusion equation is considered. Assuming no average velocity in the y and z directions, the advection term takes the same form as before:

$$\frac{\partial \bar{C}}{\partial t} + U \frac{\partial \bar{C}}{\partial x} = K_x \frac{\partial^2 \bar{C}}{\partial x^2} + K_y \frac{\partial^2 \bar{C}}{\partial y^2} + K_z \frac{\partial^2 \bar{C}}{\partial z^2} \tag{5.31}$$

In this equation, the form of the analytical solution changes depending on the boundary conditions (such as substances occurring from the surface, instantaneously occurring from the point and the existence of the ground surface). Assuming instantaneous source occurrence, which is the same as before, it is shown that the concentration follows a Gaussian distribution profile under appropriate boundary conditions as in the primary analytical solution:

$$\bar{C} = \frac{Q}{\sqrt{4\pi K_x t} \cdot \sqrt{4\pi K_y t} \cdot \sqrt{4\pi K_z t}} \exp\left(-\frac{(x - Ut)^2}{4K_x t}\right) \exp\left(-\frac{y^2}{4K_y t}\right)$$
$$\exp\left(-\frac{z^2}{4K_z t}\right) \tag{5.32}$$

This equation can be understood from the shape of (5.24) and (5.30).

Next, with approximation conditions for distance and appropriate boundary conditions for continuous occurrence, the solution is obtained as

$$\bar{C} = \frac{q}{\sqrt{4\pi K_y x}\sqrt{4\pi K_z x}} \exp\left(-\frac{y^2}{4\pi K_y (x/U)}\right) \exp\left(-\frac{z^2}{4\pi K_z (x/U)}\right) \tag{5.33}$$

From here, by integrating each case of discrete source occurrence and continuous occurrence, three-dimensional expansion ranges can be obtained as follows. The directions of the standard deviation σ_y, σ_z are estimated as follows (σ_x does not exist for continuous occurrence in the approximated expression):

$$\sigma_y = \sqrt{2K_y T}, \ \sigma_z = \sqrt{2K_z T} \tag{5.34}$$

σ_y and σ_z each indicate the expansion range of the concentration in the y direction and z direction. Here, T is the travel time, which corresponds to t, the elapsed time for instantaneous occurrence and to x/U for continuous occurrence.

Considering the value of the previous standard deviation, the diffusion in the y direction for the equation assuming continuous occurrence becomes

$$\sigma_y = \sqrt{2K_y x/U} \tag{5.35}$$

Fig. 5.5 Diffusion under
continuous occurrence
condition

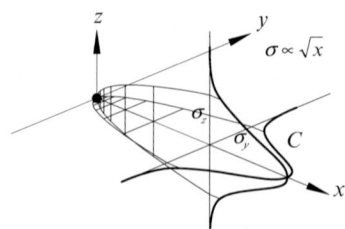

The expansion range varies with the square root against the main stream direction x. The image is shown in Fig. 5.5. In the actual measurement, the expansion range increases by an increase in the exponent to x of 0.75 to 1, and therefore, it is not appropriate. This is considered as an error that stems from assuming that the turbulent diffusion coefficient K_i is constant.

As mentioned above, K_i takes various values due to the existence of various scale vortices in turbulence. Assuming K_i to be constant means that there is only one type scale of vortex contributes to diffusion (Panofsky and Dutton 1984), which is not consistent with reality. Therefore, to obtain consistency between measurement results and this prediction formula, the increment of K must be considered. Modeling K and modeling σ means the same. Modeling σ_y and σ_z as functional types that rely on the field was attempted.

With regard to predicting the diffusion of facility exhaust, (5.31) will be examined by assuming the continuous occurrence condition. Terms involving the turbulent diffusion coefficient K_i are substituted by σ_y and σ_z. However, considering the actual phenomenon, $z = 0$ is set on the ground surface, and no absorption or occurrence of concentration at $z = 0$ are assumed. In addition, considering the diffusion from the chimneys, the pollutant is assumed to be observed at a height H from a discrete source. The emission rate is q (dimensions of $[MT^{-1}]$).

Equation 5.31 can be described as follows:

$$\bar{C} = \frac{q}{2\pi\sigma_y\sigma_z U} \exp\left(-\frac{y^2}{2\sigma_y{}^2}\right)\left\{\exp\left(-\frac{(z-H)^2}{2\sigma_z{}^2}\right) + \exp\left(-\frac{(z+H)^2}{2\sigma_z{}^2}\right)\right\} \quad (5.36)$$

Terms relating to z in the equation originate from the assumption that the substances reflect completely at the ground surface. In other words, the diffusion is expressed by locating points of virtual occurrence that are symmetric to the ground surface and summing them.

σ_y and σ_z are variables to be modeled and are expressed by means of the coordinate x in the main stream direction.

A few prediction approximations are shown below.

Sutton (1932, 1953) proposed the following equation. Constants C_y and C_z are assumed to change depending on atmospheric conditions:

$$\sigma_y{}^2 = \frac{1}{2}C_y{}^2 \cdot x^{2-n}, \quad \sigma_z{}^2 = \frac{1}{2}C_z{}^2 \cdot x^{2-n} \quad (5.37)$$

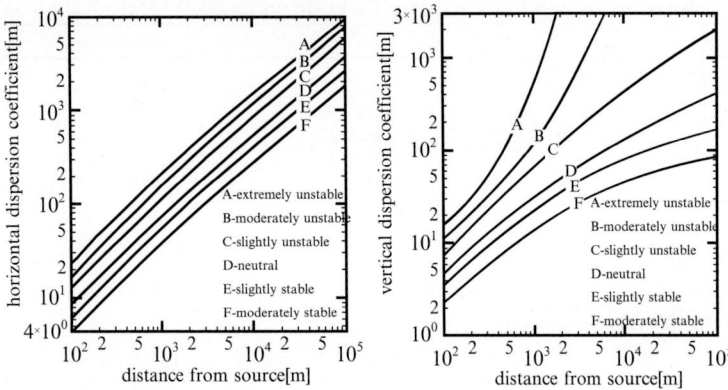

Fig. 5.6 P-G curve from Panofsky and Dutton (1984)

Table 5.1 Key to Pasquill categories from Panofsky and Dutton (1984)

	Day			Night	
Surface wind speed	Incoming solar radiation			Thinly overcast	
(at 10 m) (m/s)	Strong	Moderate	Slight	or ≥4/8 low cloud	Clear or ≤3/8 cloud
<2	A	A–B	B		
2–3	A–B	B	C	E	F
3–5	B	B–C	C	D	E
5–6	C	C–D	D	D	D
>6	C	D	D	D	D

According to Sutton, if the condition is the diffusion under a natural atmosphere and in a grassland near ground level, values $n = 0.25$, $C_y = 0.4[cm^{1/8}]$ and $C_z = 0.2[cm^{1/8}]$ are recommended (Csanady 1973). Considering the thermal environment, Pasquill (1974) classified the condition of the atmosphere regarding σ_y and σ_z. Currently, the Pasquill-Gifford diagram, a result of Pasquill, adjusted by Gifford, is the most commonly used method (Fig. 5.6).

Pasquill and others classified atmospheric condition into six classes, A–F (in some case, up to G), based on solar radiation, cloud condition, and the velocity of surface wind. They also separately determined σ_y and σ_z as a function of downwind distance x.

However, these estimated values are obtained in the field measurement. The source of smoke is located on the ground or at the height of ground, and the influence of buoyancy is not considered. The occurrence is continuous from a discrete source. The ground surface is assumed to be flat. The measurement time is approximately 3–5 min. The measurement was performed under the condition of a steady wind.

Table 5.2 Briggs revision ($10^2 < x < 10^4$[m]) from Briggs (Pasquill 1974)

Pasquill type	σ_y[m]	σ_z[m]
Open-country conditions		
A	$0.22x(1 + 0.0001x)^{-1/2}$	$0.2x$
B	$0.16x(1 + 0.0001x)^{-1/2}$	$0.12x$
C	$0.11x(1 + 0.0001x)^{-1/2}$	$0.08x(1 + 0.0002x)^{-1/2}$
D	$0.08x(1 + 0.0001x)^{-1/2}$	$0.06x(1 + 0.00015x)^{-1/2}$
E	$0.06x(1 + 0.0001x)^{-1/2}$	$0.03x(1 + 0.0003x)^{-1}$
F	$0.04x(1 + 0.0001x)^{-1/2}$	$0.016x(1 + 0.0003x)^{-1}$
Urban conditions		
A–B	$0.32x(1 + 0.0004x)^{-1/2}$	$0.24x(1 + 0.001x)^{-1/2}$
C	$0.22x(1 + 0.0004x)^{-1/2}$	$0.2x$
D	$0.16x(1 + 0.0004x)^{-1/2}$	$0.14x(1 + 0.0003x)^{-1/2}$
E–F	$0.11x(1 + 0.0004x)^{-1/2}$	$0.08x(1 + 0.00015x)^{-1/2}$

Briggs (1973), in contrast to the P-G diagram, proposed an approximation of the expansion range that considers the roughness of the urban area. The atmospheric stability was classified into six classes, the same as in the P-G diagram, and was evaluated as follows. The measurement results obtained in the experiment and the calculation results from these equations will be compared in the next chapter (Table 5.2).

5.3   Wind Tunnel Experiment Supposing the Pollutant Emission in an Urban Area

This section presents several results from a dispersion experiment inside a wind tunnel. To simulate pollutant phenomena in an urban environment, an existing urban area in Japan was recreated in a wind tunnel.

In recent years, the probability of unexpected urban pollution from either a terror attack or accidental leakage is increasing. Some well-conditioned phenomena have been revealed (point source on a flat plane, line source in the areas separating two buildings, etc.). This type of event in a highly condensed city has not been clearly studied yet. Although it is difficult to develop a general framework for dispersion in the complicated geometry of an urban city, there may be two points to take into account, which cannot be ignored.

First, the transient features of concentration have to be considered.

Second, the danger of concentration fluctuations has to be evaluated.

In past studies, the first problem was seldom mentioned. Air pollution is often thought to be caused by the exhaust from chimneys or from traffic in street canyons. The effect of these pollution sources on humans is often evaluated by an annual cumulative dosage. In contrast, in modern situations, the pollution is emitted, and its dispersion is unpredictable. Even after we detect the origin of the source, we have to determine the best-suited evacuation route. Thus, the time scale of the problem is quite short, i.e., how long the pollutant takes to arrive a distance after the start of emission and how long the pollutant takes to vanish after emission has stopped, and may be of great concern.

The second problem represents the danger of an intermittent high concentration. Because of the nonlinearity in the transport equation of momentum, the fraction of scalar also shows a fluctuation in the flow field. It is plausible that the danger should be evaluated by the cumulative dosage. The average concentration may also effectively estimate the value; however, this may lead to ignorance about the second nature of the danger, i.e., the pedestrian may intake a fatal dose in a moment even if the average concentration at the location he was located was estimated to be quite low. Thus, we have to take the probability density function of concentration into consideration.

To simulate the conditions, a wind tunnel experiment was conducted. The sources and measurements were conducted on a model of an existing area in Tokyo, Japan. The instrument with a high-frequency response was utilized for the concentration measurements. The instantaneous response of the concentration was measured under an unsteady discharging condition. The data were reduced to show the transient features, and the stochastic properties of the concentration fluctuations were evaluated.

5.3.1 Experimental Setup

Wind tunnel: The wind tunnel at the Institute of Industrial Science at the University of Tokyo was utilized. It has a cross section with a 2.2-m width and a 1.8-m height, as well as a 16.47-m inlet length. The thermal condition in the tunnel is neutral. 1.5-m-high vortex generators were set downstream of the contraction flow area. The roughness was aligned upstream of the experimental area.

Measurement: An approach flow was measured with a constant temperature anemometry (CTA:Dantec). The probe was 55P11. The concentration was measured with a flame ionization detector (FID:Technica). The sampling tube was 0.4 m in length with a 100-Hz response frequency. The velocity sampling was conducted at 1,000 Hz, while that for the concentration measurement was conducted at 100 Hz.

Targeted area: The scale of the recreated model was 1/500. The Iidabashi district, which consists of both medium high-rise office buildings and a residential area, was targeted in the experiment.

Dimensionalization: The concentration, time, and frequency will be shown in dimensionalized form:

$$C^* = CuL^2/Q, \tag{5.38}$$

Fig. 5.7 Sketch of the urban model

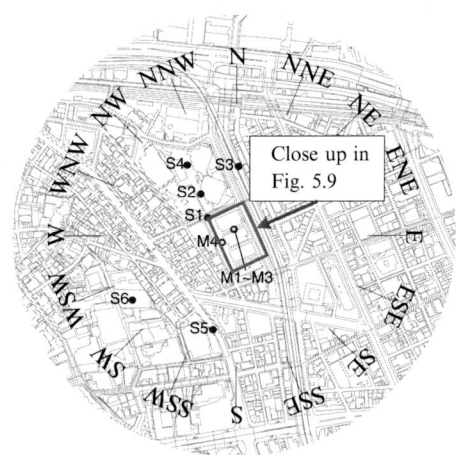

$$t^* = tu/L, \qquad (5.39)$$

$$f^* = fL/u, \qquad (5.40)$$

where C (m^3/m^3) is the instantaneous concentration, t (s) is the time, f (s^{-1}) is the frequency, L (0.125 m) is the referential height, Q (1×10^{-6} m^3/s) is the dosage of tracer, and u (1.14 m/s) is the velocity of the approach flow at the referential height.

Measurement and source points: In the model, several sources and measurement points were set. Figure 5.7 illustrates the recreated urban model. The measurement points were plotted as M1–M4. The source points were plotted as S1–S6. S1, S2, and S4 are located in between the high-rise buildings. Figure 5.8 shows a close-up view of the source points. S3 is located on a highway, S5 is located on a trunk road, and S6 is located on a school ground. The measurement points were located in the space in between the high-rise buildings (Fig. 5.9). From the ground level, M1, M2, and M3 are located at heights of 2.5, 30, and 60 m, respectively, in terms of real scale. The location of the measurements corresponds to the "void" described by Bu et al. (2009). M4 is located in front of an office building entrance.

Approach flow: The approach flow was recreated to fit a 1/4 power law. Figure 5.10 shows the approach flow and power spectrum at a 200-mm height in wind tunnel scale. The power spectrum shows a good fit to the Karman spectrum. The longitudinal turbulent length scale was almost 0.5 m. The horizontal turbulent length scale was 100 mm.

Tracer emissions: The inner diameter of the tracer's source was 6×10^{-3} m. Ethylene was utilized as the tracer, and the amount of emission was 1×10^{-6} m^3/s.

Unsteady source discharging: Figure 5.11 shows a sketch of the experimental conditions. A mass flow controller (Ohkura) was utilized for the tracer's constant mass flow. The tracer's route was separated in two ways. One was connected to the experimental system and another to the exhaust. The tracer was emitted in a discretized way by switching the route with a pressure valve system. After a step

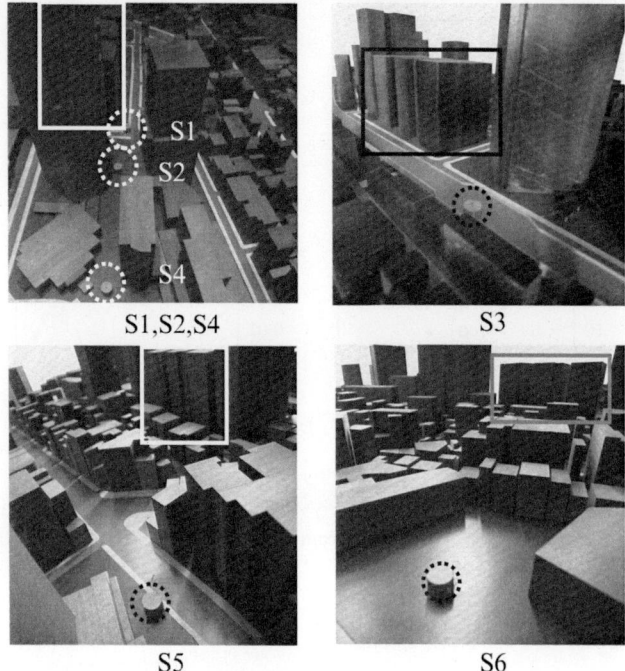

Fig. 5.8 Close-up of the pollutant sources

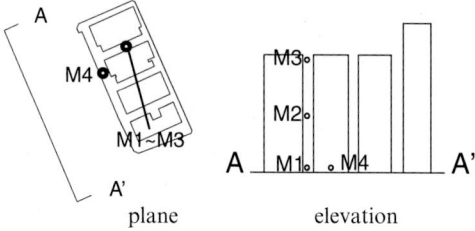

Fig. 5.9 Measurement locations

up in emission, the tracer was continuously emitted for a sufficient time. After the average concentration field became constant, the emission was stopped suddenly. The concentration was measured at both the source point and the measurement point. The data were recorded with an A-D recorder (NR-2000: Keyence). The time of these two outputs was synchronized. The time the tracer starts emission was detected from the source point's dataset. Frequencies of more than 12.5 Hz of digital data were cut.

The time history of the concentration may be different even if the acquisition condition is unchanged. To acquire the average tendencies of the concentration's transient features, data acquisition at every source and measurement condition was

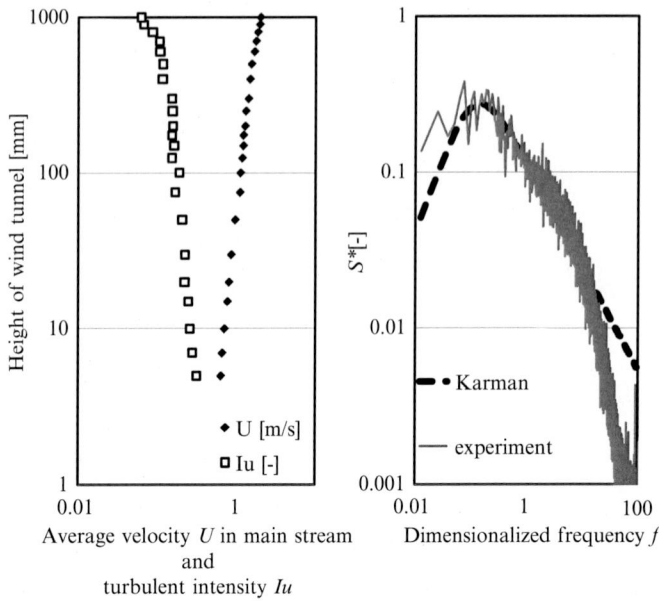

Fig. 5.10 Approach flow and power spectrum

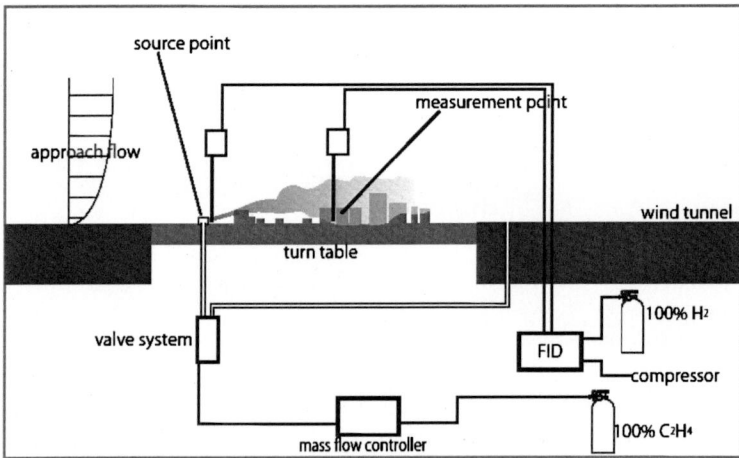

Fig. 5.11 Sketch of the experiment

conducted 32 times. To understand the average tendency of the concentration response, the data were averaged together.

Measurement cases: The measurements were conducted under the conditions described in Table 5.3. In the table, S2-SW-M1 means the source point is S1, the wind direction is SW, and the measurement point is M1. The measurements of

Table 5.3 Measurement cases

Source point	Wind direction	Measurement point
S1	NNW	M4
	NW	M1, M2, M3, M4
	WNW	M1, M2, M4
	W	M1, M2, M4
	WSW	M1
S2	NNW	M4
	NW	M1, M2, M3, M4
S3	NNE	M1, M2, M4
	N	M1, M2, M4
	NNW	M1, M2
S5	SSW	M4
	S	M1, M2, M3, M4
S6	SW	M1, M2, M4

unsteady discharging condition were carried out only in the cases that the average concentrations of 1 min measurement exceed 30 ppm. In the cases of source point S4, the average concentrations did not exceed the threshold in any measurements.

Higher-order moment and fitting of the probability density function: Equations 5.41–5.46 show expressions for the higher-order moment:

$$\mu = E(C) = \int CP(C)dC = C_{mean} \tag{5.41}$$

$$\mu_2 = E(C - \mu)^2 = \int (C - C_{mean})^2 \, P(C)dC = C_{r.m.s} \tag{5.42}$$

$$\mu_3 = E(C - \mu)^3 = \int (C - C_{mean})^3 P(C)dC \tag{5.43}$$

$$\mu_4 = E(C - \mu)^4 = \int (C - C_{mean})^4 P(C)dC \tag{5.44}$$

$$SKEW = \frac{\mu_3}{\mu_2^{3/2}} \tag{5.45}$$

$$KURT = \frac{\mu_4}{\mu_2^2} - 3 \tag{5.46}$$

Equations 5.47, 5.48 and 5.49 show the fitted probability density function (Csanady 1973; Hanna et al. 1984).

Gaussian distribution

$$p(C) = \frac{1}{\sqrt{2\pi}C_{r.m.s}} \exp\left\{-\frac{(C - C_{mean})^2}{2C_{r.m.s}^2}\right\} \tag{5.47}$$

Log-normal distribution

$$p(C) = \frac{1}{\sqrt{2\pi}\sigma_c C} \exp\left\{ -\frac{\{\log(C/n_c)\}^2}{2\sigma_c^2} \right\}$$

$$n_c = \frac{C_{mean}}{\sqrt{1 + (C_{r.m.s}/C_{mean})^2}}$$

$$\sigma_c = \sqrt{2\log(C_{mean}/n_c)} \qquad (5.48)$$

Exponential distribution

$$p(C) = \frac{I^2}{C_{mean}} \exp\left(-I\frac{C}{C_{mean}}\right) + (1 - I)\delta(C = 0)$$

$$I = \frac{2}{(C_{r.m.s}/C_{mean})^2 + 1} \qquad (5.49)$$

where $P(C)$ means the probability density function of the concentration, and I means the intermittency rate.

5.3.2 Results

Figure 5.12 shows the time series of the concentration after the start of tracer emission. The bold lines show the average of the 32 measurements. The lines, which impinge on the bold line, show the t-test's two-sided 95% confidence interval. Figure 5.13 shows the average time the tracer takes to arrive at the measurement locations. Table 5.4 shows the average concentration as well as its standard deviation, normalized third-order moment (skewness), and normalized fourth-order moment (kurtosis). Figure 5.14 shows the concentration's probability density function. The three types of functions (Eqs. 5.48–5.50) were fitted to the experimental results.

Case S1-NW: Faster arrival times were observed in comparison to another case. The measurements near the ground level (M1, M4) showed especially faster arrival times. As the measurements got higher, the arrival times were delayed. The distance between the measurement points and the source point may cause this result. The average concentration and its standard deviation showed large values at lower measurements. As the measurements got higher, the values became smaller. In this case, every measurement point showed maximum values during each measurement. The skewness was around 1 to 2, while the kurtosis varied from 3 to 6. These values are not necessarily large in comparison to the other cases. The pdf in M1, M2, and M4 showed a good fit to the log-normal distribution, while M3 showed a good fit to the exponential distribution.

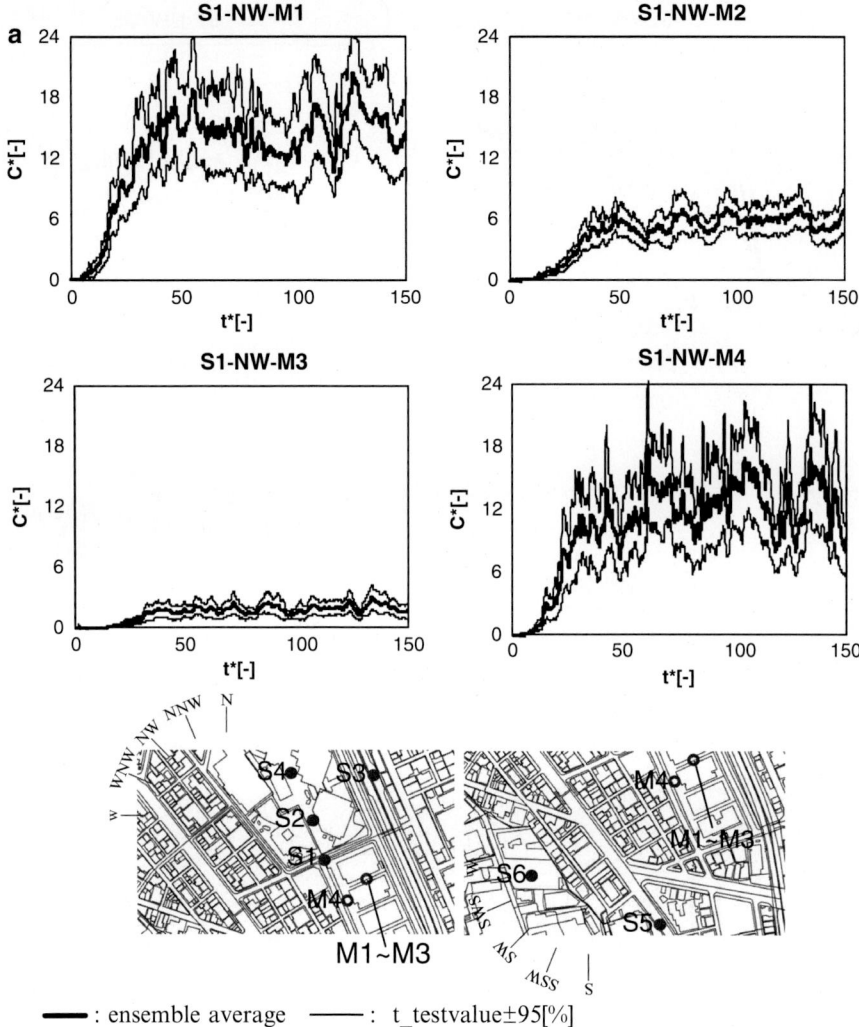

Fig. 5.12 Transient responses at each location

Case S2-NW: M1 showed a faster response followed by M2 and then M3. Although the order is the same as in S1-NW, the arrival time showed somewhat of a delay in comparison to the S1-NW case. The distance between the measurements and the source might cause this tendency.

The average concentration and its standard deviation were larger for M1 and M2. As the measurements got higher, these properties took smaller values. The skewness was around 2 to 4, while the kurtosis varied from 4 to 24. The skewness and kurtosis in M1 and M4 were both maximum at these measurement points and conditions. The shapes of the pdf in M1, M2, and M4 showed better fits to log-normal distributions, while M3 was better fit by an exponential-like distribution.

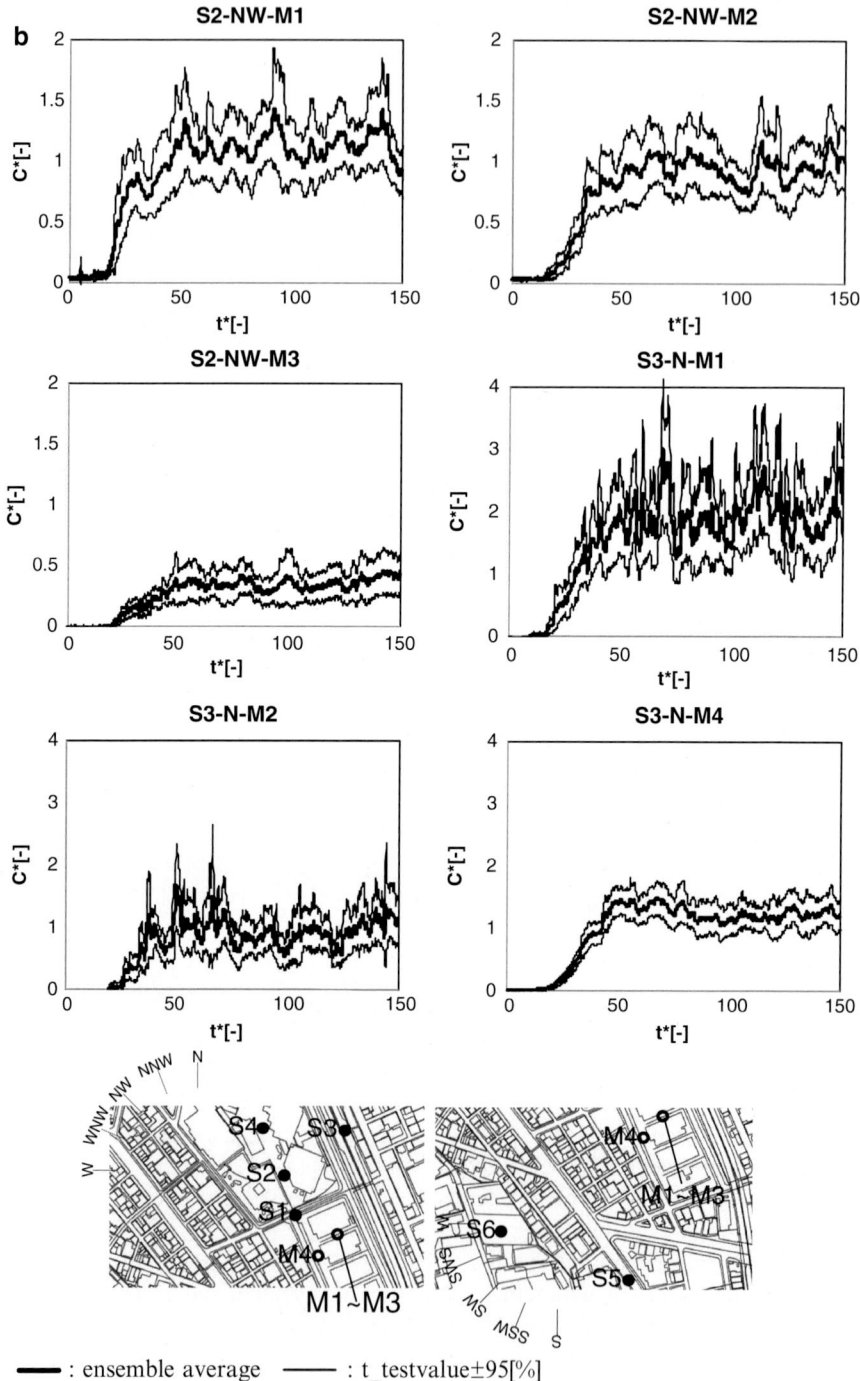

Fig. 5.12 (continued)

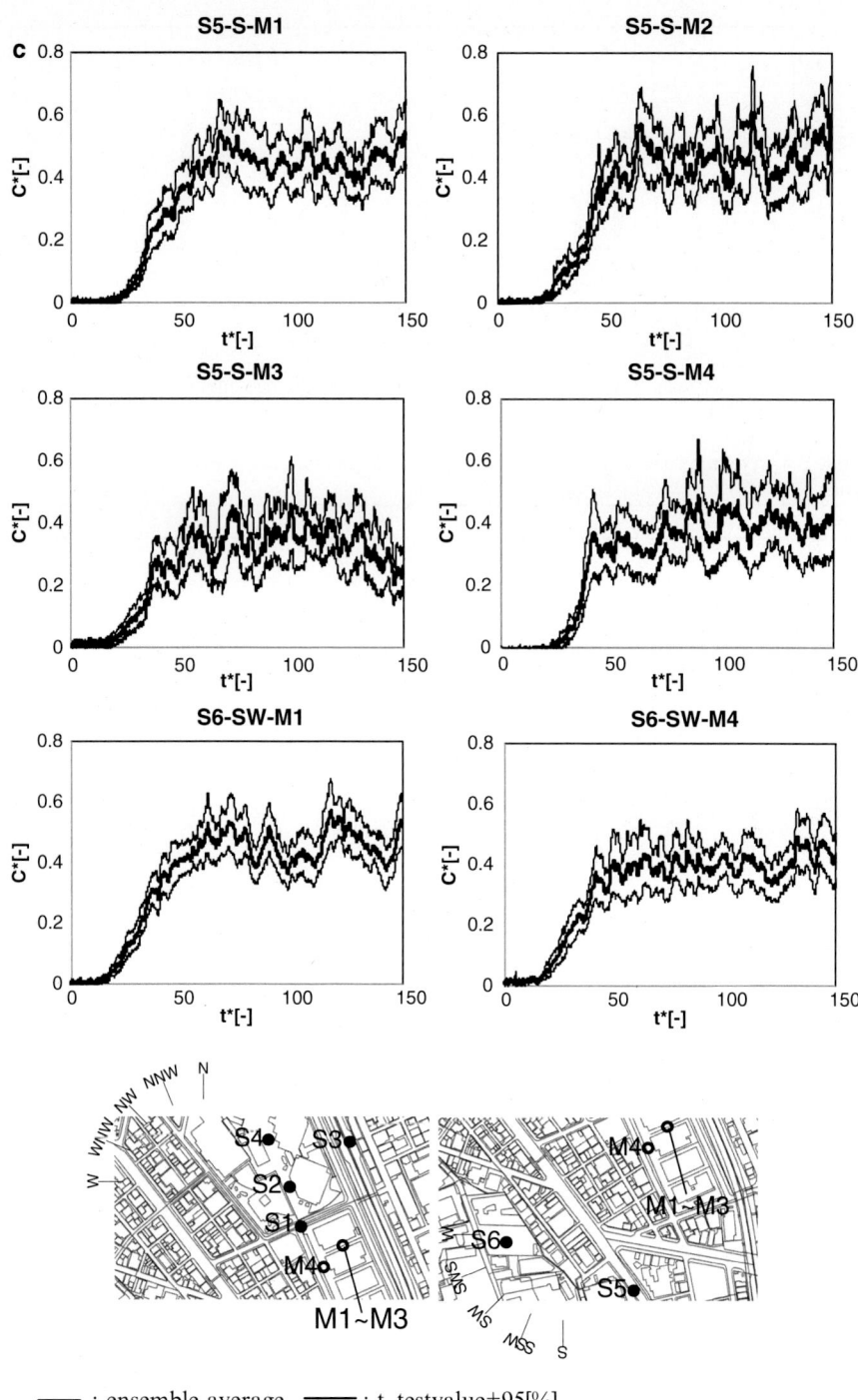

: ensemble average   ———— : t_testvalue±95[%]

Fig. 5.12 (continued)

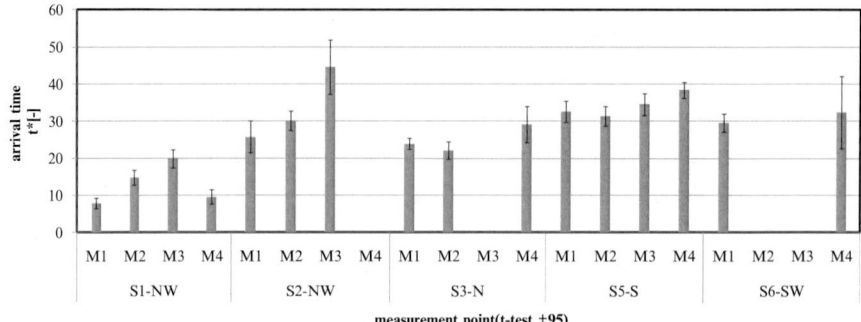

Fig. 5.13 Arrival times t^* and their t test values at $\pm 95\%$ confidence interval

Table 5.4 Average concentration (ave), standard deviation (std), skewness (skew), and kurtosis (kurt)

Cases		Ave	Std	Skew	Kurt
S1-NW	M1	15.540	11.492	1.433	2.933
	M2	5.940	4.394	1.466	3.239
	M3	1.967	2.355	1.951	4.400
	M4	12.445	11.662	2.036	5.797
S2-NW	M1	1.179	0.821	2.497	11.371
	M2	0.916	0.618	1.971	7.220
	M3	0.401	0.445	1.940	4.828
	M4	1.206	1.066	3.789	24.609
S3-N	M1	1.951	1.736	2.014	7.200
	M2	0.946	1.045	3.089	19.964
	M3	–	–	–	–
	M4	1.319	0.717	0.980	1.835
S5-S	M1	0.491	0.264	0.628	0.969
	M2	0.497	0.307	1.314	3.507
	M3	0.374	0.291	1.378	2.590
	M4	0.423	0.313	1.193	3.319
S6-SW	M1	0.472	0.207	0.749	0.834
	M2	0.375	0.252	1.293	2.820
	M3	–	–	–	–
	M4	0.422	0.223	0.948	1.106

Case S3-N: The arrival time showed almost the same value in M1 and M2. This might be because the source point was located in a highway, which had a height between M1 and M2. The response in M4 was delayed. This might be because the measurement point was located on the back side of a high-rise building. The result showed a more delayed arrival time in comparison to S1-NW, but some of the results showed faster arrival times than S2-NW. These results may be because the source was located in a highway, the location without obstacles.

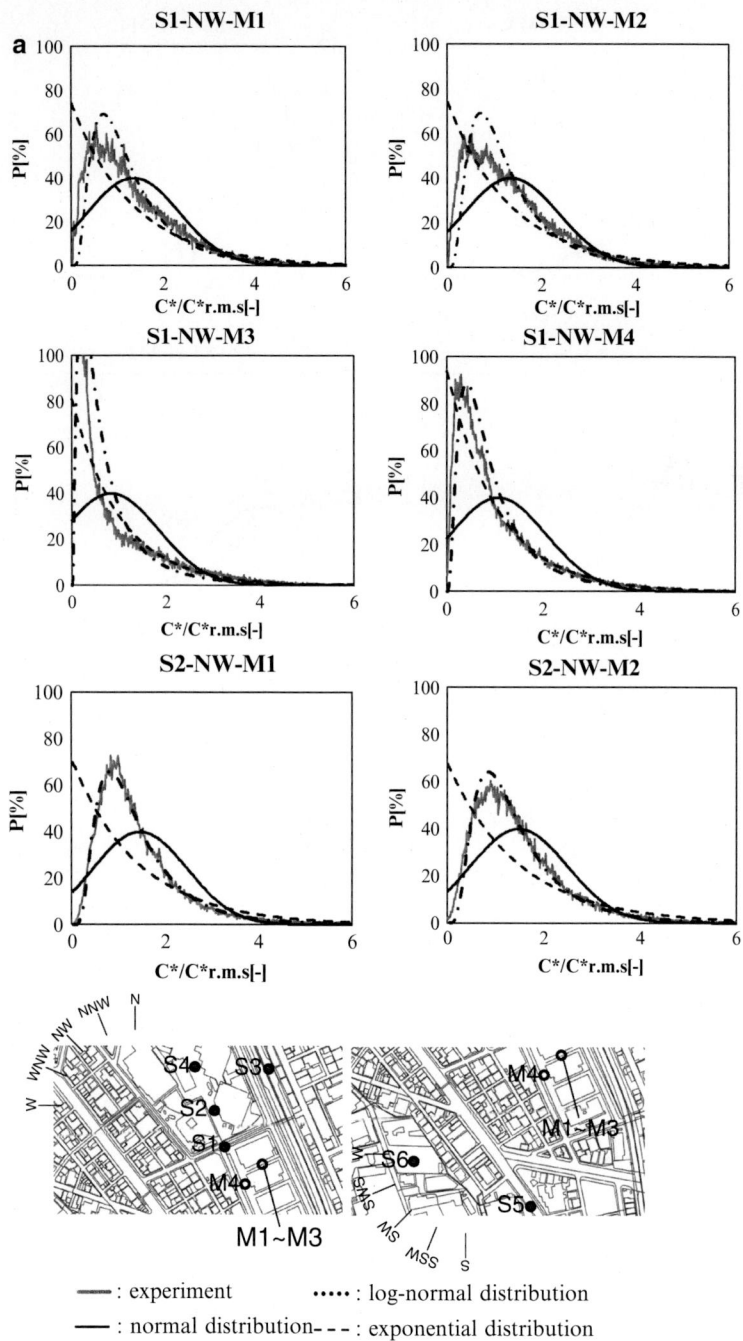

Fig. 5.14 Probability density functions

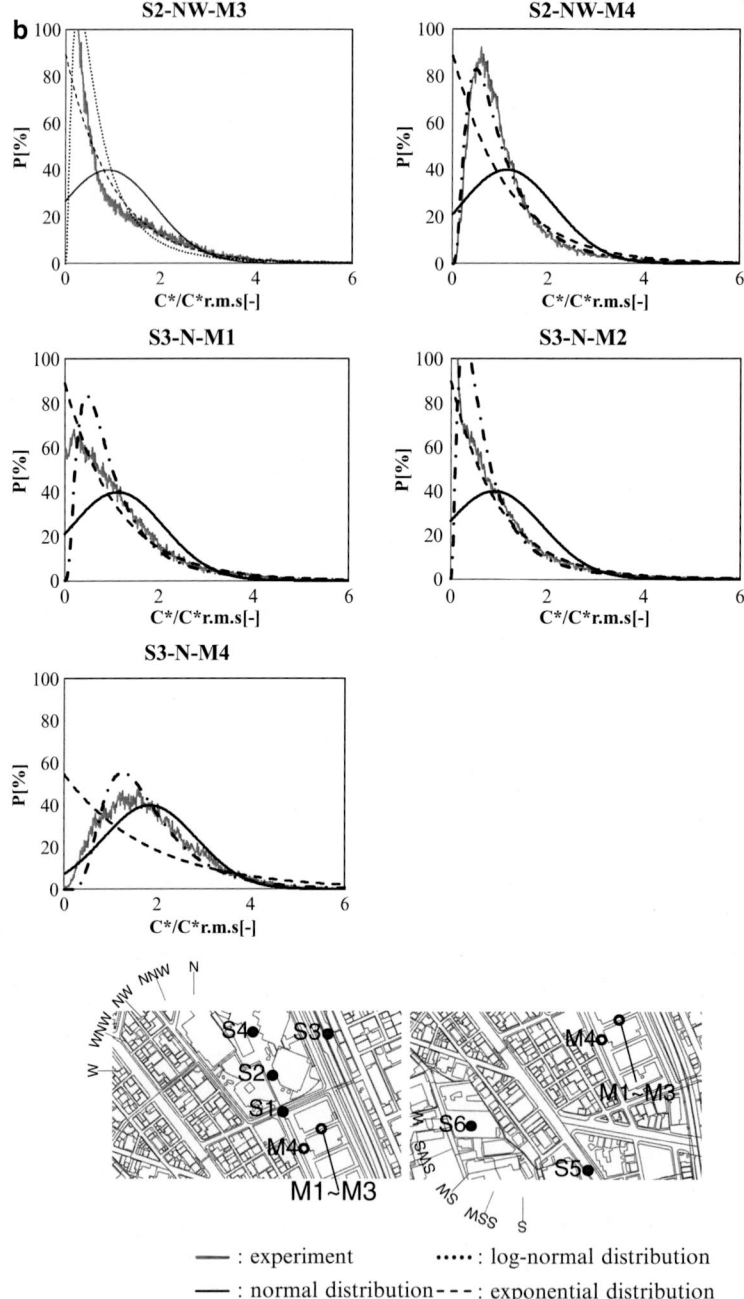

Fig. 5.14 (continued)

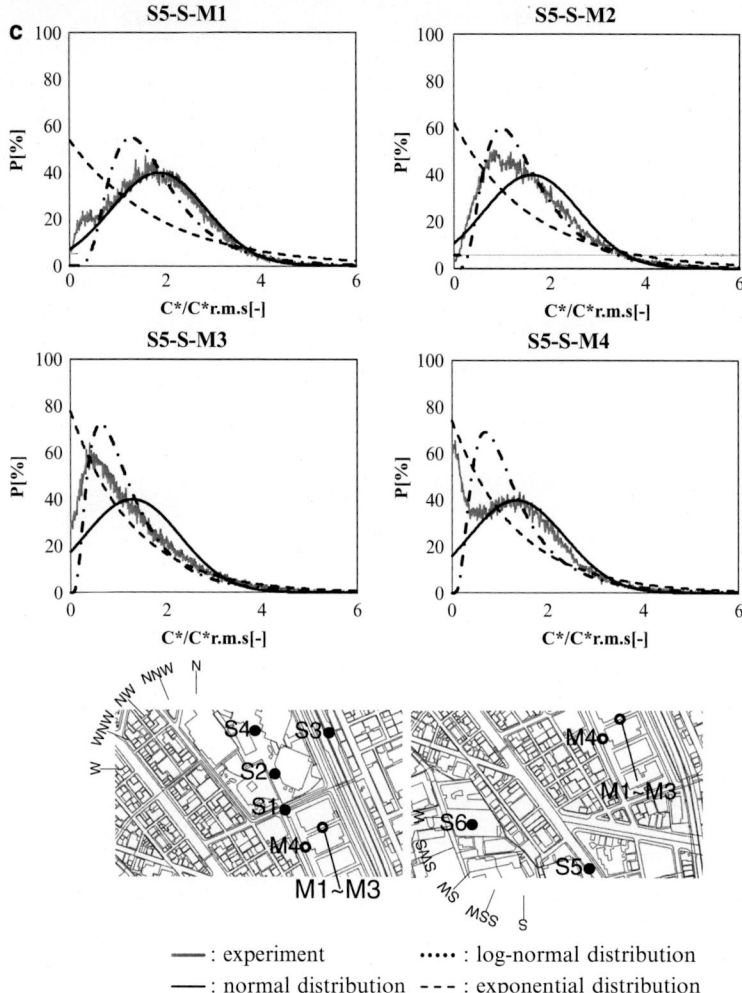

Fig. 5.14 (continued)

The average concentration and its standard deviation had a smaller value than S1-NW; however, some cases took a larger value than the S2-NW case. M4 had a smaller standard deviation. The alignment of the buildings strongly affected the concentration fluctuation. The skewness was around 1 to 3, while the kurtosis varied from 1 to 20. M2 had a maximum value, while M1 showed a second maximum value at each measurement point in this case. M4 had a lower value both in skewness and kurtosis. The probability density function for M1 and M2 had an exponential-like distribution. The tracer might seldom arrive at this point. M4 had a log-normal-like distribution.

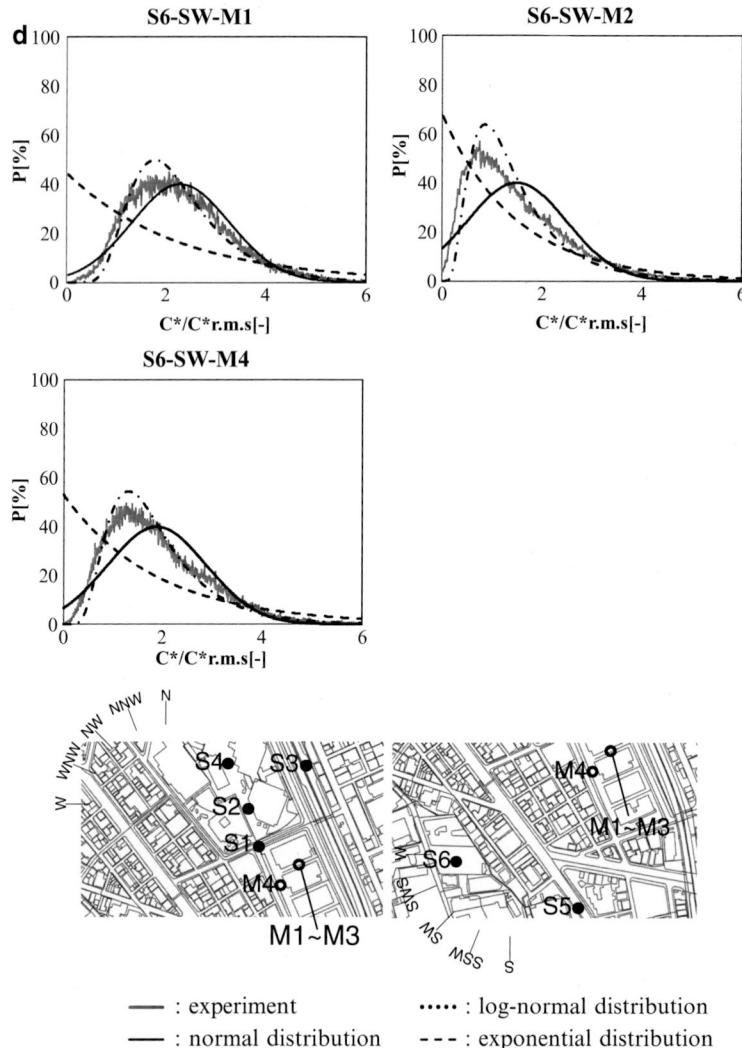

Fig. 5.14 (continued)

Case S5-S: The arrival time for M2 was the shortest followed by M1, M3, and M4. All of the measurements showed a delayed arrival time, and the difference among them was small.

The average concentration and its standard deviation showed comparatively small values. The difference among the measurements became small. The skewness varied from 0.5 to 1.5, while the kurtosis was roughly 1 to 3.5. Compared to the other cases, the values were small, the shapes of the pdf were not distorted negatively or positively, and its shape did not spread well. The distance between the measurement and the source point may enhance turbulent diffusion. As a result, the concentration fluctuation may have decayed. The shape of the pdf showed a normal-like distribution for M1 and a log-normal-like distribution for M2. For M4, the profile seemed to be an intermediate shape between an exponential distribution and a log-normal distribution.

Case S6-SW: The dataset in M2 was insufficient due to noise. The arrival times for M1 and M4 showed a similar tendency. The average concentration for M3 did not exceed 30 ppm. The transient response was not measured.

The average concentration took small vaule and the difference of it between the measurements is also small. The skewness was around 1, while the kurtosis varied from 1 to 3. Along with S5-S, they both showed a lower value. The distortion of the pdf was not strong, and it did not spread well. Both of the pdfs showed a log-normal-like distribution.

5.3.3   Comparison with the Diffusion Approximation and Experiment

The result obtained in the experiment in the previous section and the predicted concentration calculated using the diffusion approximation in Sect. 5.2 are compared in this section. The average concentration will be calculated from an approximation based on the P-G diagram and the Briggs equation. The Briggs equation is the equation proposed to add the roughness of the city to the P-G diagram equation as the evaluation target (see 5.2.3 for the Briggs equation). The Briggs equation is also classified by A–F atmospheric conditions. The calculation is based on D (neutral) in this document.

Figure 5.15 shows the expansion range of the P-G diagram and the Briggs equation. Compared to the P-G diagram, the Briggs equation determines a large expansion range in both the y and z directions. In addition, the expansion range in the horizontal direction is set to be larger than that in the vertical direction.

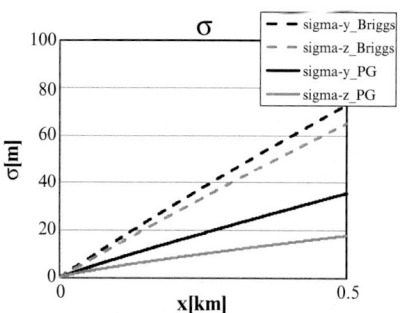

Fig. 5.15 Expansion range of P-G diagram and Briggs equation

Table 5.5 Geographical conditions in each case

Cases	M1–M3		M4	
	x[m]	y[m]	x[m]	y[m]
S1-NW	59.7	22.5	61.8	17.0
S2-NW	109.7	4.7	111.7	44.6
S3-N	143.1	10.0	172.8	36.7
S5-S	226.9	47.0	197.1	20.2
S6-SW	273.7	47.9	233.8	50.0

Fig. 5.16 The x and y directions in each case

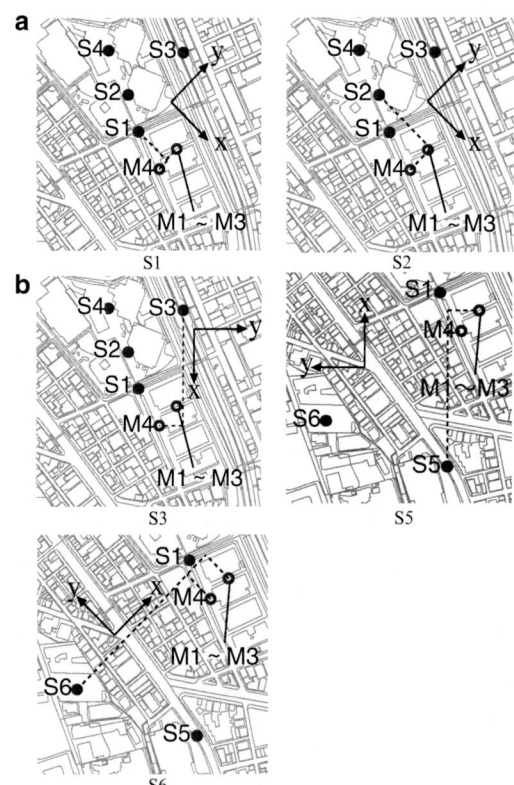

Table 5.5 shows the distance from the point of occurrence of the pollution in the downwind direction (x direction) and horizontal direction to the point of occurrence (y direction) in each case. Figure 5.16 shows the geometric condition of measurement points and source points. In the diagram, the x coordinate shows the mainstream wind direction, and the y coordinate shows the horizontal direction. Table 5.6 shows the experiment results and calculation results obtained by the prediction equation.

S1-NW (Case where the distance between the occurrence point and measurement point is the shortest): The approximation yielded smaller values than the experiment results. Between the measurement point and occurrence point, there are large

Table 5.6 Average concentration comparison

Cases		Wind tunnel	P-G	Briggs
S1-NW	M1	15.540	0.002	0.887
	M2	5.940	0.000	0.002
	M3	1.967	0.000	0.000
	M4	12.445	0.197	2.953
S2-NW	M1	1.179	19.956	3.909
	M2	0.916	0.000	0.624
	M3	0.401	0.000	0.002
	M4	1.206	0.000	0.199
S3-N	M1	1.951	1.872	1.796
	M2	0.946	0.116	0.879
	M3	–	0.000	0.075
	M4	1.319	0.067	0.622
S5-S	M1	0.491	0.159	0.426
	M2	0.497	0.001	0.281
	M3	0.374	0.000	0.080
	M4	0.423	3.581	1.014
S6-SW	M1	0.472	0.319	0.385
	M2	0.375	0.007	0.274
	M3	–	0.000	0.117
	M4	0.422	0.116	0.381

buildings, and they narrow the transport path of concentration. Three-dimensional average wind speeds exist in the narrowed transport path. In this case, the It might cause to restrict the property of turbulent diffusion in a horizontal way and the tracer is stagnated in the canyon of the street. That might be the reason why high concentration in the measurements were observed. However, these tendency happening in the canyon of the urban city is not reflected in the approximation for the diffusion prediction .

Near the originating point, because the expansion ranges in the horizontal and vertical directions were not as large as the y, z coordinates of M1–M4, the concentration was 0 at M2 and M3. On the other hand, in the experiment, because of the buildings, the turbulent diffusion in the vertical direction was promoted and took recognizable value.

S2-NW (Case where the distance between the occurrence point and measurement point is the second shortest): The approximation showed a higher value than the experiment at M1 because M1 is located almost on the 0 point of y coordinate. Then, the result for the prediction was lower than the experiment value at measurement points M2 to M4 because the diffusion in the horizontal and vertical directions was underestimated in the approximation, and therefore, the concentration took the recognizable value only in the region where the length of y direction is almost 0. The P-G diagram overestimates the concentration at measurement point M3, where y is extremely close to 0; however, the concentration remains 0 at the other points. On the other hand, in the Briggs equation, this tendency is restrained. A slight vertical diffusion is shown, and relatively good approximation can be observed at M2. However, the horizontal diffusion is not sufficient, and a huge error is seen in the result at M4, where the y coordinate is large.

S3-N (Case where the distance between occurrence point and measurement point is the third shortest): In this case, the occurrence point is located at the height of 12.5 [m] in actual scale. The result of the approximation using the P-G diagram does not show the correct result at M2 or M3. However, the result at the measurement point on the ground surface M1 was fairly accurate. The approximation using the Briggs equation showed a good result in the vertical diffusion as well. However, there was an error between the experiment and approximation at M4 because the diffusion property was very different than the approximation as the measurement point was located behind the building.

S5-S (Case where the distance between the occurrence point and measurement point is the fourth shortest): The P-G diagram still does not reproduce the horizontal and vertical diffusion sufficiently. Because the measurement point M4 has a small y coordinate, the approximation based on the P-G diagram shows a high concentration. However, it remained 0 at the points M2 and M3. The result shows that the diffusion in the height direction is remarkably low. The result calculated using the Briggs equation is similar to the experimental result for the vertical direction diffusion as well.

S6-SW (Case where the distance between the occurrence point and measurement point is the fifth shortest): Similar to the S5-S case, the P-G diagram did not reproduce the diffusion in the horizontal and vertical directions sufficiently. A nonzero concentration was observed at the measurement points near the ground surface (M1 and M4); however, the value remained fairly close to 0 at points M2 and M3. Thus, the influence of diffusion in the height direction is not considered. The results calculated using the Briggs equation is similar to the experimental results for the vertical direction diffusion as well.

Nomenclature for Sect. 5.3

C	Concentration of tracer	$[m^3/m^3]$
C_{mean}	Average concentration of tracer	$[m^3/m^3]$
$C_{r.m.s}$	Root mean square of concentration	$[m^3/m^3]$
C^*	Nondimensionalized concentration	$[-]$
$C^*_{r.m.s}$	Root mean square of nondimensionalized concentration	$[-]$
C^*_{mean}	Nondimensionalized average concentration of tracer	$[-]$
f	Frequency	$[s^{-1}]$
f^*	Nondimensionalized frequency fL/u	$[-]$
I	Intermittency	$[-]$

Iu	Turbulent intensity	$[-]$
$KURT$	Kurtosis	$[-]$
L	Referential height	$[m]$
μ	Average value	$[m^3/m^3]$
μ_2	Second-order central moment	$[(m^3/m^3)^2]$
μ_3	Third-order central moment	$[(m^3/m^3)^3]$
μ_4	Fourth-order central moment	$[(m^3/m^3)^4]$
P	Probability density function	$[\%]$
Q	Mass of tracer emission	$[m^3/s]$
S	Spectrum power	$[m^2/s]$
$S*$	Nondimensionalized power $f \cdot S/\sigma^2$	$[-]$
$SKEW$	Skewness	$[-]$
σ	Root mean square of velocity	$[m/s]$
t	Time	$[s]$
$t*$	Dimensionalized time	$[-]$
u	Referential velocity	$[m/s]$
U	Average velocity of main stream in wind tunnel	$[m/s]$

5.4   Conclusions

The prediction method for the problems regarding the pollutant diffusion was described. Then, based on various ground surface roughnesses and classifications of the atmospheric thermal environment, the tuning method of the diffusion range was described with a prediction method based on the Gaussian diffusion model.

In addition, the wind tunnel experiment was performed to investigate the concentration response under nonsteady temporal substance occurrence conditions because a new kind of pollutant diffusion problem was noted. The focus was on the time that the concentration diffuses from the occurrence point of pollution in the high-density urban area with a complex town shape. Measurement points at each location were located in the wind tunnel model created based on the actual city, and multiple measurements were performed under the same conditions. An ensemble average was performed on the time history data of concentration. These data were described with the city characteristics and distance between the occurrence point and measurement point. A comparison was made between the average concentration obtained and the diffusion prediction model. The standardized high-order moment values around the center were indicated along with the profile of the probability density distribution. The experimental result assuming realistic conditions was explained. However, a deeper general consideration regarding the concentration fluctuation is required in the future.

Acknowledgment The present research in Sect. 5.3 is financially supported by the sponsorship of MEXT (Ministry of Education, Culture, Sports, Science and Technology) of Japan from 2007 to 2009 and is conducted in corporation with Mitsubishi Heavy Industries.

References

Briggs GA (1973) Diffusion estimates for small emissions. ATDL Contribution No.79 Atmospheric Turbulence and Diffusion Laboratory, Oak Ridge, Tenn

Bu Z, Kato S, Ishida Y, Huang H (2009) New criteria for assessing local wind environment at pedestrian level based on exceedance probability analysis. Build Environ 44(7):1501–1508

Csanady GT (1973) Turbulent diffusion in the environment. Reidel Publishing Company, Dordrecht

Hanna SR (1984) The exponential probability density function and concentration fluctuations in smoke plumes. Bound Layer Meteorol 29:361–375

Panofsky HA, Dutton JA (1984) Atmospheric turbulence-models and methods for engineering applications. Wiley-Interscience, New York

Pasquill FA (1974) Atmospheric diffusion, 2nd edn. Halstead Press-Wiley, New York

Sutton OG (1932) A theory of eddy diffusion in the atmosphere. Proc R Soc Lond A 135:143–165. doi:10.1098/rspa.1932.0025

Sutton OG (1953) Micrometeorology. McGraw-Hill Book Co., New York

Part II
Criteria for Assessing Breeze Environment

Chapter 6
Legal Regulations for Urban Ventilation

Kyosuke Hiyama and Shinsuke Kato

Abstract The performance of urban ventilation correlates strongly with air quality and thermal comfort in urban areas. A good living environment that includes such features is a citizens' right that needs to be secured for the people living in urban areas. Proper legal regulation is necessary to secure this right. In this chapter, we discuss the existing legal regulations concerning urban ventilation. There are two concepts for legal regulations: prescriptive-based code and performance-based code. We discuss the Building Standard Law in Japan as an example of a prescriptive-based code and the Air Ventilation Assessment System (AVAS) in Hong Kong as a candidate for future performance-based codes. It is thought that legal regulation is not currently sufficient to ensure the aforementioned right, because the correlation between the regulations and urban ventilation is not clear. We also refer to studies on how to properly evaluate ventilation performance, and we then discuss the concepts of "horizontal ventilation paths" and "vertical ventilation paths."

Keywords Building code • Prescriptive-based code • Performance-based code • Ventilation path • Urban ventilation

6.1 Introduction

When buildings in urban areas are dense, they block ventilation. This results in lower transport efficiency of pollution by the wind, which causes the degradation of air quality. In addition, the transport of waste heat by the wind and the experience of a cool breeze that is enjoyed by humans are also both degraded. It is expected that the comfort of an open-air thermal environment during the summer season would

K. Hiyama (✉) • S. Kato
Institute of Industrial Science, The University of Tokyo, 4-6-1, Komaba,
Meguro-ku, Tokyo 1538505, Japan
e-mail: hiyama@iis.u-tokyo.ac.jp1; kato@iis.u-tokyo.ac.jp2

S. Kato and K. Hiyama (eds.), *Ventilating Cities: Air-flow Criteria for Healthy and Comfortable Urban Living*, Springer Geography, DOI 10.1007/978-94-007-2771-7_6, © Springer Science+Business Media B.V. 2012

thus be degraded. A good living environment, including good air quality and thermal comfort, should be a right and needs to be secured for the people living in urban areas. However, in urban areas where a large mass of people live, an attempt by an individual to optimize his or her right in utilizing his or her own land may violate the rights of others who utilize adjacent land. Thus, it becomes necessary to limit individual rights in order to minimize this conflict of interest and to ensure the impartiality of the rights that land users should enjoy.

Legal regulations that are needed to ensure proper ventilation are discussed in this chapter. First, legal regulations and guidelines from Japan and Hong Kong, China, are provided as examples. In these locations, degraded ventilation because of extreme overpopulation is a conspicuous problem, and these examples provide an overview of the conditions for a limitation on the right to ventilation. In this discussion, the concepts of "prescriptive-based codes" and "performance-based codes" in legal regulations will be explained. Next, how to evaluate the performance of ventilation, as it is required to establish such legal regulations, will be discussed based on a review of previous ventilation studies. Finally, the concepts of "horizontal ventilation paths" and "vertical ventilation paths" will be explained.

6.2 Prescriptive-Based Codes and Performance-Based Codes

To maintain a high level of environmental quality for the users of a building, relevant regulations guaranteeing that buildings achieve proper environmental performance should be established. Such regulations are generally categorized into prescriptive-based codes and performance-based codes. A prescriptive-based code is a specific rule regarding the specifications of a building. A performance-based code, on the other hand, only specifies the standard of performance to be secured by the regulation without using exact specifications; thus, any buildings that fulfill the regulations are permitted, regardless of the specifications. For a light environment, where a regulation is established to ensure a certain level of daylight, a prescriptive-based code specifies the area ratio of the windows, whereas a performance-based code allows any design to be used, provided that the daylight factor is in the acceptable range, which is verified based on a daylight calculation. Consequently, performance-based codes require more work in the calculations than prescriptive-based codes, but they allow more tolerance in the design.

In general, the safety and comfort of the living environment as a product of ventilation are achieved by prescriptive-based codes that limit the density of the population in each urban block. Nevertheless, questions have arisen about the relationship between the control of population density in urban areas by prescriptive-based codes and the improvement of the living environment to date; thus, a movement to establish performance-based codes for ventilation is currently active. To review the current regulation standards for ventilation, Japan is first used as an example to explain regulation standards based on prescriptive-based codes. The implementation of performance-based codes is then explained using the guidelines for heat islands in Japan and the guidelines in Hong Kong, China, as examples.

6.2.1   Applications of Prescriptive-Based Codes

6.2.1.1   The Building Standard Law in Japan

The Building Standard Law in Japan (The Building Center of Japan 2009) is generally divided into the categories "Summary Law," "Institutions Law," and "Substantial Law." The "Summary Law" defines the purpose or terms of the Building Standard Law. The "Institutions Law" covers processes, including procedures and penal regulations. The "Substantial Law" specifies the structure of a building. The right to safety and comfort, which should be secured for the people living in the building, is protected by this "Substantial Law." This law is based on the minimum functional property of the building, so the safety and comfort of users could be ensured by a city where the mass of buildings maintains an adequate environment.

This "Substantial Law" is divided into the "individual control system" and the "bulk control system." The individual control system is the standard that is applicable to buildings in any of the areas in Japan. It includes regulations on structural strength, fireproofing and evacuation, safety and health, and facilities for construction. It specifies the structure of buildings irrespective of adjacent buildings or the condition of the site. When this standard is observed, the security and comfort of the building are permanently ensured for that particular building. Specifically, this standard defines the structural strength and other specifications required to protect the lives and property of building users, as well as the specifications and equipment required to ensure that features such as proper lighting capacity or ventilation capacity, which are mandatory to maintain the health of the users, exist inside the rooms.

In contrast, the bulk control system includes the regulations that are applicable only to the specific urban areas designated by each prefecture (city planning area/quasi city planning area). It includes regulations on roads, land-use zones (zones that regulate the use of buildings for possible construction), shape control, fireproof areas, and aesthetic districts. Shape control regulates the floor area ratio, building coverage ratio, setback distance (i.e., the distance between the boundary of the site and the outer wall), and the shape of the building, including height. The rights of the building users in the whole area to security and comfort, including ventilation, as discussed in this chapter, could be achieved by the Substantial Law that controls the shape of buildings and establishes appropriate urban blocks.

The simplest idea for ensuring ventilation is to use this shape control to restrict the density of urban blocks, which limits the building coverage ratio or floor area ratio. Then, an excessive concentration of buildings is avoided by including vacant areas. Parks and roads also contribute to restricting the density of urban blocks. Table 6.1 shows an example of typical zoning as well as the shape control in each zone. In areas assigned as residential zones, a lower building coverage ratio is specified, which increases the vacant area with the intention of creating a living environment with proper ventilation and lighting. In areas assigned as commercial zones, a higher building coverage ratio is specified, which maximizes the

Table 6.1 Land-use zone and shape control (floor area ratio/building coverage ratio[b])

Land-use zone[a]	Category 1 exclusively low-rise residential zone	Category 1 mid/ high-rise-oriented residential zone	Commercial zone	Industrial zone
Floor area ratio (%)	50–200	100–300	200–400	200–400
Building coverage ratio (%)	30–60	60	80	30–60
Setback distance of outer wall (m)	1, 1.5	–	–	–
Absolute height limit (m)	10, 12	–	–	–

[a]Other than this table, 8 other land-use zones are specified
[b]In addition to this table, the "setback-line limit" to limit the height of buildings using an imaginary line drawn from the adjacent land or the "shadow limit" to limit the shade on the day of the winter solstice is specified per each use zone, for the purpose of securing ventilation and lighting of the neighborhood

economics of the land. However, the living environment is predicted to degrade in this case. According to the Building Standard Law, commercial zones are allowed the highest building coverage ratio. In areas assigned as exclusively industrial, the building coverage ratio is similar to residential zones. Primary consideration is given to safety hazards, such as fire, as well as to improving the environment.

However, these current regulations were not established based on a detailed verification of the relationship between the regulations and the environment. Thus, it is not certain to what extent safety or environmental standards are ensured by these regulations. As for ventilation issues, previous studies (Kubota et al. 2000, 2002; Yoshie et al. 2008) have reported that a correlation was observed between the gross land ratio (ratio of the building area against a total area in which parks, roads, and vacant lands are included) and the wind velocity in residential zones (Fig. 6.1).

However, it has also been pointed out that it is difficult to evaluate the extent to which wind velocity reduces pollution within residential zones. It is difficult to provide clear evidence of why only the prescriptive-based codes can secure ventilation. In addition, urban blocks that were formed before the establishment of the Building Standard Law have many existing buildings that do not qualify and that fail to satisfy the shape control standard. In these cases, the difficulty in satisfying these standards and the fact that the existing conditions may function as a preventive factor against the reconstruction of old buildings are recognized problems. As a result, numerous urban blocks in Japanese cities consist of a set of buildings that provide insufficient safety or comfort. The Building Standard Law intends to secure these things for users, but updating such cases is quite challenging.

In recent years, a program has been carried out by the legal committee of the Architectural Institute of Japan. They plan to amend the current bulk control system from prescriptive-based code-oriented to performance-based code-oriented. Whereas a prescriptive-based code is based on the building coverage ratio or floor area ratio and is specified uniformly in certain areas, a performance-based code specifies only that target standards be met. To find the best solution for the

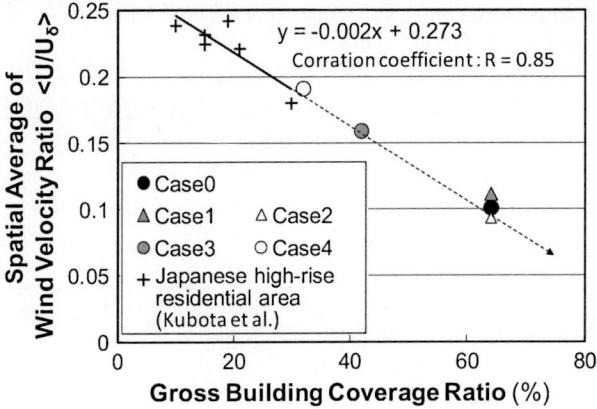

Fig. 6.1 Relationship between gross building coverage ratio and spatial average of wind velocity ratios (Yoshie et al. 2008)

bulk control system, these problems with the Building Standard Law are being closely observed. A group of several areas of urban blocks is used as the subject of the study, and the environmental performance of certain urban areas as a whole is being studied, evaluated, or verified.

6.2.2   Applications of Performance-Based Codes

6.2.2.1   Criteria for Assessing a Strong Wind Environment

Although the concerns are different from those referred to in this book, performance-based criteria for a strong wind environment are widely used today. In this section, we present the acceptable criteria proposed by Murakami et al. (1983, 1986) for assessing a strong wind environment, which are widely used in Japan today. These criteria are frequently used in court to judge whether or not there are wind-induced issues associated with construction. With this background, these criteria are often used to evaluate the effect of construction on the wind environment. The criteria establish an acceptable wind speed level and evaluate the occurrence of wind-induced problems based on the frequency at which the daily maximum gust wind speed occurs. The wind-induced problems are evaluated based on the frequency of occurrences of the daily maximum gust wind speed because pedestrians may not recognize the daily change in the average wind velocity, but they will certainly notice days with wind that is strong enough to cause wind-induced problems. There are several other studies that discuss assessments of the wind environment and refer to the concept of an acceptable frequency for the occurrence of strong winds. Hunt et al. (1976) studied the correlation between wind

Table 6.2 Acceptable criteria for a wind environment based on the frequency of occurrences of the daily maximum gust wind speed (Murakami et al. 1983)

Class	Effect of strong wind	Applicable areas (example)	Strong wind assessment level and acceptable not-to-exceed frequency (at a height of 1.5 m)		
			Daily maximum gust wind speed (m/s)		
			10	15	20
			Daily maximum mean wind speed (m/s)		
			10/GF	15/GF	20/GF
1	Areas used for purposes most susceptible to wind effects	Shopping street in residential area; outdoor restaurant	10 (37 days/ year)	0.9 (3 days/ year)	0.08 (0.3 days/ year)
2	Areas used for purposes not too susceptible to wind effects	Residential area; park	22 (80)	2.6 (13)	0.6 (2)
3	Areas used for purposes least susceptible to wind effects	Office street	35 (128)	7 (26)	1.5 (5)

When these criteria are applied, values of wind-tunnel experimental maximum mean wind speed can be used as an indicator of wind speed. The maximum mean wind speed can be found by converting the gust wind speed using a gust factor

GF = gust factor (height 1.5 m, averaging time 2–3 s). Area where wind speeds are particularly high (1.6–2.5); typical city area (2.0–2.5)

$GF = f/f_g$, where f is 10-min mean wind speed and f_g is maximum 3-s gust

speed and acceptable criteria using a wind tunnel experiment, and Davenport and Isyumov (Isyumov 1975) and Melbourne (Melbourne 1978) studied criteria that were based on the concept of an acceptable frequency for the occurrence of strong winds. Among these studies, the work of Murakami et al. is unique in that they conducted a monitoring survey of human consciousness for 2 years and developed the criteria based on the results.

Table 6.2 shows the acceptable criteria for a wind environment based on the frequency of occurrences of the daily maximum gust speed. The locations for the evaluation are divided into three classes based on whether or not the location is susceptible to the wind. For example, at a class 1 location, the frequency of occurrences of wind speeds above 10 m/s must be below 10% per annum, or 37 days per annum; this criterion dictates an acceptable wind environment for a residence. If we evaluate the wind environment using the daily maximum mean wind speed, for example, in case a wind tunnel experiment is used for the evaluation, we can use a gust factor to convert the daily maximum mean into the gust wind speed. As we have mentioned previously, the criteria are based on the results of monitoring surveys from residences. Hence, the criteria connect directly to absolute figures that indicate whether the wind conditions cause wind-induced problems. In this context, the criteria are performance based.

6.2.2.2 CASBEE-HI: Comprehensive Assessment System for Built Environment Efficiency for Heat Island Relaxation

In recent years, the phenomenon in which the temperature in urban areas is significantly higher than in the surrounding suburbs (a heat island) has become the subject of public concern. The development of a heat island is viewed as an issue that violates the rights of residents because it degrades the living environment, especially in the summer. Tokyo, Japan, is one of the most notable cities that suffer from the heat island effect (Ministry of the Environment and Japan 2002; Chen et al. 2009). The number of days in a year in which the lowest temperature exceeds 25°C has increased threefold when compared with the 1920s (Ministry of the Environment and Japan 2010). This detrimental shift toward a hotter environment has the effect of increasing the number of heat stroke patients. Controlling the shape of the buildings that compose the city for the purpose of reducing the heat island effect is an urgent task that will ensure the right of the residents to enjoy healthy lives.

In 2001, Japan's unique environmental performance evaluation system, "CASBEE" (Japan Sustainable Building 2010a; Murakami et al. 2004), was developed under the initiative of the Ministry of Land, Infrastructure and Transport. CASBEE is similar to BRE Environmental Assessment Method (BREEAM) established in the UK by the Building Research Establishment (BRE) (Yates et al. 1998) or Leadership in Energy and Environmental Design (LEED) developed by US Green Building Council in the USA (USGBC 2009; Crawley and Aho 1999; Fenner and Ryce 2008a, b); the purpose of CASBEE is to add value to buildings by rating their environmental performance and to serve in local governments as the standard of control for buildings. One feature of this system is its method for calculating the value of an evaluation. The Building Environmental Efficiency (BEE) as a value of the evaluation is obtained using Eq. 6.1 below:

$$BEE = \frac{Q}{L}, \tag{6.1}$$

where Q is the quality, or the total value of environmental performance, and L is the load, or the total evaluation value of the environmental load.

A noticeable feature of this equation is that when the environmental performance is high and the environmental load is low, it is possible to receive a high evaluation.

This system is still being improved upon today, and an expansion tool to be utilized as a countermeasure to the aforementioned heat island effect, CABEE-HI (CASBEE for Heat Island Relaxation) (Japan Sustainable Building 2010b; Oguro et al. 2008), is also being developed. This system evaluates the extent of the contribution of each building to reducing the heat island effect. This system

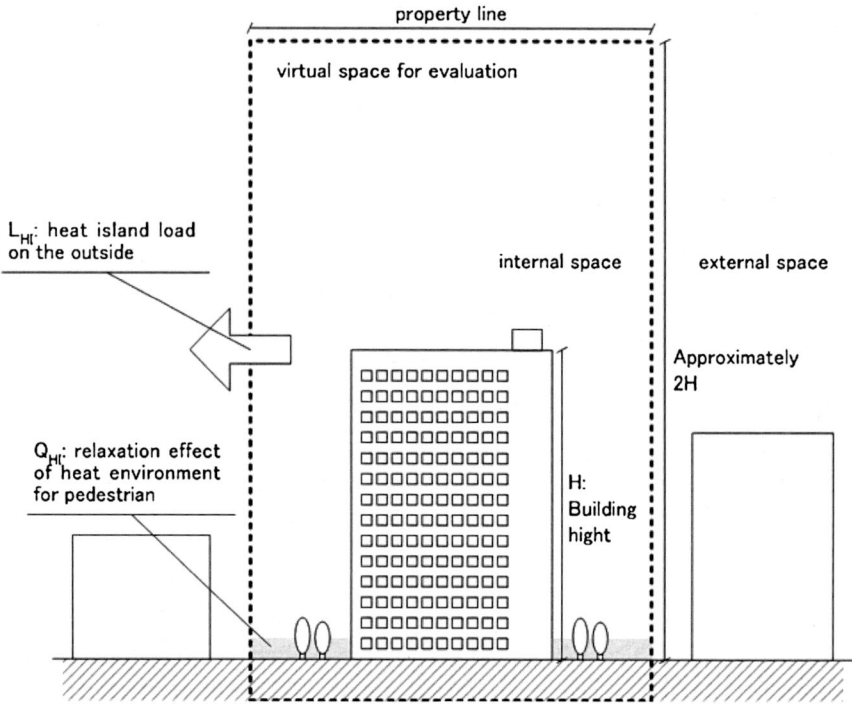

Fig. 6.2 Model of CASBEE-HI: the virtual space for evaluation and the concept of Q_{HI} and L_{HI}

improves on Eq. 6.1 and calculates the evaluation value of BEE_{HI} (Building Environmental Efficiency for Heat Island Relaxation) using Eq. 6.2:

$$BEE_{HI} = \frac{Q_{HI}}{L_{HI}}, \tag{6.2}$$

where Q_{HI} is the quality, or the relaxation effect of the heat environment within the virtual closed space, and L_{HI} is the heat island load on the outside of the virtual closed space. Figure 6.2 shows the concept of Q_{HI} and L_{HI}.

The active ventilation of a city is attracting attention as a method for reducing the heat island effect. With this method, cool air is moved from the suburbs to the urban areas and chills the urban areas. Thus, in CASBEE-HI, items that are related to ventilation are reflected in the rating system, as listed in Table 6.3.

As the linking of the condition "reduction of the projected area toward the prevailing direction of wind" in L_{HI}-1 (2) to a high score indicates, this system incorporates the horizontal winds in urban areas. The system is devised so that buildings in the upwind area will earn a high rating by not obstructing the ventilation of buildings in the downwind area. Although CASBEE is a rating system that evaluates buildings using performance-based criteria, items that earn the credit are

Table 6.3 Rating system items related to urban ventilation

Main heading	Subheading	Detailed heading
Ventilation	Q_{HI}-1 Induce wind into the pedestrian region in the target site to relieve the heat environment	1. Review the condition of wind in areas adjacent to the target site to plan the arrangement and shape of buildings, to enable wind induced into the pedestrian region and others
		2. Provide vacant lands with turf, grass, or bushes to create a ventilation path
	L_{HI}-1 Pay attention to ventilation toward the downwind area to relieve the thermal effect of the area outside the target site	1. When planning the arrangement and shape of buildings, consideration should be given to not block the wind toward the downwind area
		2. The projected area of the buildings toward the prevailing wind direction during the summer time will be reduced
		3. Height and shape of the buildings and the distance between building blocks will be taken into account to avoid blocking the wind

still prescriptive-based in the sense that the evaluation is based on the projected area of a building or the distance between building blocks. CASBEE is used to evaluate the environmental performance of buildings that are already functioning in major self-governing bodies such as Osaka City or Nagoya City. Because it is obligatory to submit a score for the construction of buildings, CASBEE functions as a guideline for construction within these self-governing bodies. In the future, CASBEE-HI is also expected to function as a guideline for the composition of urban blocks. Much consideration based on research has been given to the method of grading these evaluation indexes, and it is still being improved upon today.

6.2.2.3 Air Ventilation Assessment System (AVAS) in Hong Kong

The Government of Hong Kong has required an assessment of urban ventilation to be conducted for all major government projects since December 2006 (Ng and Ren 2009). To conduct these assessments, the Government of Hong Kong has developed and adopted the Air Ventilation Assessment (AVA) (Ng 2009), which evaluates low-velocity wind. Because the goal of the AVA is to allow wind to flow from the suburbs into urban areas, the focus in Hong Kong is on constructing buildings in urban areas that do not disturb this wind. VR_w (wind velocity ratio) is used as an indicator of this evaluation, as shown in Eq. 6.3:

$$VR_w = \frac{V_p}{V_\infty}, \tag{6.3}$$

where V_∞ is the wind velocity in the upper air in a location that is not affected by buildings and V_p is the wind velocity in the residential space (2 m above the ground). Sixteen directions are accounted for in VR_w, where weight is assigned in accordance with the probability for each wind direction. Although it is possible to determine the wind velocity using CFD (computational fluid dynamics), a wind tunnel test is recommended. The VR_w will be reported as the SVR_w (site spatial average wind velocity ratio) and the LVR_w (local spatial average wind velocity ratio). The SVR_w is the average value of VR_w obtained from measurement points that are aligned at the boundary of the development area at intervals of 10–50 m. The SVR_w is used to evaluate the effect of development on the wind environment at the boundary of the developing area. The LVR_w is the average value of VR_w obtained from measurement points that are arranged in the area surrounding the development area (i.e., the area enclosed by a line drawn along the boundary of the development area at a distance equal to the height of the highest building within the development area) at intervals of 200–300 m². The LVR_w is used to evaluate the effect of development on the wind environment in the local area. The VR_w is used as an indicator to evaluate the extent to which people staying in the target urban block enjoy the wind that is directed to them from the suburbs. A higher indicator corresponds to a smaller disturbance by buildings that violate the right to enjoy wind; thus, the evaluation is that the effect of the development on the wind environment is lower. This system of assessment has already been used for several projects, and the reports are available on the web pages of the Planning Department and the Government of the Hong Kong Special Administrative Region (2010). Because the present aim of the system is to merely report the results of the assessment and not to control the shape of the buildings, it is expected that the system will also function as a performance-based code when a minimum standard is specified in these guidelines.

6.3   Ventilation Performance

When regulating the ventilation in urban areas using a performance-based code, the definition of the ventilation performance is a controversial point. Both the CASBEE-HI in Japan and the AVA in Hong Kong intend to prevent buildings from being constructed in developing areas that block wind flow from the suburbs of the developing area. Thus, it can be said that they focus on maintaining the horizontal wind environment. These intentions may have their roots in the idea of the "ventilation path," which was introduced in Stuttgart, Germany (Ichinose 1993; Narita 2006). Stuttgart is an industrial city that is focused on the automobile industry. With the expansion of industry, a heat island problem and serious air pollution issues emerged because the city is in a conical basin where the atmosphere tends to stagnate. Thus, the urban planning manager, in cooperation with climatologists, developed a plan for the city whereby fresh, cool air from the hills surrounding the urban area can flow into the urban area. For this purpose, they imposed restrictions prohibiting the construction of certain buildings and limiting the height of buildings, and they developed parks and green lands as well. As a result, the carbon monoxide concentration was reduced to a level much lower than the standard values in Germany, which also led to a reduction in temperature in the urban areas. This ventilation resulted in a significant reduction in both the atmospheric pollution and the heat island effect. Based on this success in Stuttgart, the goal today in urban planning is to find a place where cool air is produced (cool spots) and then arrange a ventilation path for the purpose of reducing the heat island effect and atmospheric pollution. Oguro et al. used CFD analysis to optimize the horizontal ventilation path from a river on a residential estate (Oguro et al. 2002). Horikoshi et al. demonstrated the cooling effect of a sea breeze traveling up a canal (2001), and the results have been connected to future city planning designs for the city of Nagoya in Japan. In cities like Tokyo or Hong Kong that are open to the sea, it is theorized that introducing a large volume of cool air produced in the sky over the sea to replace the air in the urban area would significantly improve the urban environment. However, when comparing the case of the city of Stuttgart, where cool air is introduced from the hills, with Tokyo's plan, which introduces cool air from the sea, it is clear that the scale and mechanism of the phenomena are different. Consequently, careful consideration is required when introducing a similar urban planning scheme to a different city to determine whether it will have the same effect. First, the ventilation path in Stuttgart is based on the horizontal flow of wind, where an upstream and a downstream direction are created. The cool air produced in the cool spot is first consumed in the upstream area, which results in the attenuation of the chilling effect that is provided by the "ventilation path" plan in proportion to the distance from the cool spot. As such, it is expected that a plan to utilize the cool air from the sea as the cool spot may provide the desired result along the coastline, but it will be difficult to achieve the same benefit in an inland area. In addition, when considering the purification of the pollution emitted in urban areas, as the fresh air in the upstream area is polluted because of the "ventilation path" mechanism, the same purification effect from ventilation cannot be expected to occur in the downstream area. Consequently, as an alternative to the conventional "horizontal ventilation path," a plan is being developed to introduce fresh, cool air from the sky above an urban area. This air will chill and purify the urban area's atmosphere through a "vertical ventilation path," where the wind passing in the sky above the urban area is directed into the urban area (Yoshie 2008).

The thickness of the layer of cool air above the city of Stuttgart is approximately 10 m maximum, but the cool layer of air produced over the sea can be up to 1–1.5 km thick (Yoshie 2008). When this layer is only 10 m thick, the presence of buildings in the upwind area may hinder the delivery of cool air to the buildings in the downwind area. On the other hand, when the layer is 1–1.5 km thick, the streams of cool air can pass through above the buildings, regardless of the shape of the buildings in the upwind area. Thus, even without the existence of a ventilation path from the cool spot, it is possible to introduce cool air into the urban area by redirecting the wind from the skies high above it. This vertical introduction of wind is defined as the "vertical ventilation path" as opposed to the "horizontal ventilation

Fig. 6.3 Image of the horizontal ventilation path

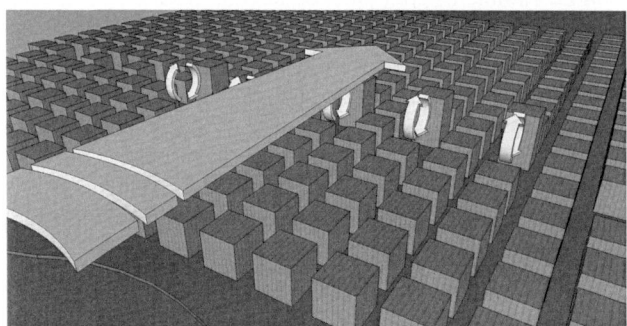

Fig. 6.4 Image of the vertical ventilation path

path." Figure 6.3 shows an image of the "horizontal ventilation path." Figure 6.4 shows an image of the "vertical ventilation path." With a "horizontal ventilation path," high-rise buildings can block the wind stream and prevent the ventilation effect in urban areas, whereas with a "vertical ventilation path," high-rise buildings that are blocking the wind stream function as components of the vertical ventilation path to pull the fresh, cool air down. For the "horizontal ventilation path," because the cool air is consumed and polluted in the upwind area, the extent of its application is limited. On the other hand, the plan to improve the ventilation of urban areas with the "vertical ventilation path" utilizes a large volume of wind from the sky, so the extent of its application is greater. However, even for wind from the sky, there is a view that states that the cool air delivered by the wind is also consumed in the upwind area near the cool spot, similar to the "horizontal windward area," and this point should not be ignored (Mochida 2009).

6.4 A Performance-Based Code Concept

The aforementioned idea to introduce fresh, cool air from the sky to an urban area through a "vertical ventilation path" refers to the replacement of air that is contaminated with waste heat and pollution that is produced in the urban area with air from higher in the sky for the purpose of reducing this waste heat and pollution. In other words, it is identical to "ventilating the urban area." The efficiency of such ventilation varies depending on the freshness of the air or the amount of heat in the air in the sky. However, if you consider that the air in the sky is much cleaner than the air in urban areas, it is possible to universally evaluate the ventilation in urban areas using the scale for ventilation efficiency (SVE), which provides the efficiency of pollution discharge and waste heat removal (Kato and Huang 2009). When performance is specified by an established regulation, this universal method becomes essential.

Using this idea (Kato 2007), a scale to enable a universal evaluation of ventilation in urban areas is discussed in the next chapter and beyond, based on the following principles:

1. Ventilation refers to the continuous flow of air. Because it is difficult to evaluate this on a point scale, it should be evaluated within a continuous area. Namely, ventilation shall be measured and evaluated in an area with a defined extent. For example, the amount of wind associated with proper ventilation is difficult to define using a point scale, but is easy to define using an area. Once this area is defined, the amount of wind passing through the area can also be defined. The physical quantity of the wind that is evaluated using points shall also serve as a scale for rating the ventilation in terms of the evaluation of distribution properties and such within the area. The simplest scale with which to define the distribution properties is the average space value.
2. Wind is a phenomenon of probability and statistics, so ventilation is also evaluated as a statistical value within a certain period, such as in each of the four seasons or in a year.
3. The strength of the wind, which corresponds to the amount of ventilation, is generally defined by the wind velocity. However, the wind velocity is a vector quantity and includes the direction of the wind. Thus, to define the strength of the wind, it is preferable to use the scalar quantity, which is the kinetic energy of the wind, rather than the vector quantity.
4. The kinetic energy that defines the strength of the wind does not directly correspond to the transport performance, which is the capability to remove waste heat and polluted air. Ventilation is evaluated using the scale for ventilation efficiency (SVE), which defines the efficiency of the discharge of pollution and waste heat, or the efficiency of transportation by the wind.

6.5 Conclusions

In this chapter, we discussed the codes that are used to optimize urban ventilation. First, we introduced existing codes/guidelines, such as the Building Standard Law and CASBEE-HI in Japan and AVAS in Hong Kong. Through these examples, we explained the difference between prescriptive-based codes/guidelines and performance-based codes/guidelines. However, there is currently no performance-based code/guideline to assist urban planners with optimizing ventilation performance, because no study has produced a method for universally evaluating ventilation performance. Currently, researchers have been studying the "ventilation path" and developing the concept of a "vertical ventilation path." The concept of a "vertical ventilation path" refers to the introduction of fresh, cool air from the sky into an urban area. The purpose of this concept is to ventilate waste heat and polluted air from an urban area using the air higher in the sky. If we assume that the air in the sky is cleaner than the air in urban areas, it is possible to universally evaluate the ventilation in urban areas using the scale for ventilation efficiency (SVE). In the next chapter, we discuss a method for evaluating the urban ventilation efficiency that is based on the SVE.

References

Chen H, Ooka R, Huang H, Tsuchiya T (2009) Study on mitigation measures for outdoor thermal environment on present urban blocks in Tokyo using coupled simulation. Build Environ 44:2290–2299. doi:10.1016/j.buildenv.2009.03.012

Crawley D, Aho I (1999) Building environmental assessment methods: applications and development trends. Build Res Inf 27:300–308

Fenner RA, Ryce T (2008a) A comparative analysis of two building rating systems. Part 1: Evaluation. Proc Inst Civ Eng Eng Sustain 161:55–63. doi:10.1680/ensu.2008.161.1.55

Fenner RA, Ryce T (2008b) A comparative analysis of two building rating systems. Part 1: Evaluation. Proc Inst Civ Eng Eng Sustain 161:65–70. doi:10.1680/ensu.2008.161.1.65

Hashimoto T, Funahashi K, Horikoshi T (2001) The cooling effect of going-up sea breeze on the urban thermal environment: the case of Horikawa canal and Shin-Horikawa canal in Nagoya. J Archit Plann Environ Eng Trans AIJ 545:65–70 (in Japanese)

Hunt JCR, Poulton EC, Mumford JC (1976) The effects of wind on people; new criteria based on wind tunnel experiments. Build Environ 15–28. doi:10.1016/0360-1323(76), 90015-9

Ichinose T (1993) "The Wind Channel" in Stuttgart, Germany. Tenki 40:691–693 (in Japanese)

Isyumov N, Davenport AG (1975) The ground level wind environment in built up area Proceedings of the 4th International Conference on Wind effects on buildings and structures, London, 403–422

Japan Sustainable Building Consortium (2010a) Comprehensive assessment system for building environmental efficiency CASBEE for new construction—Technical Manual 2008 Edition. Institute for Building Environment and Energy Conservation (IBEC), Tokyo

Japan Sustainable Building Consortium (2010b) Comprehensive assessment system for building environmental efficiency CASBEE for heat island relaxation. Institute for Building Environment and Energy Conservation (IBEC), Tokyo, Japan

Kato S (2007) How the ventilation through and over urban blocks is measured and evaluated. Wind Eng JAWE 32:421–423 (in Japanese)

Kato S, Huang H (2009) Ventilation efficiency of void space surrounded by buildings with wind blowing over built-up urban area. J Wind Eng Ind Aerodyn 97:358–367. doi:10.1016/j. jweia.2009.05.003

Kubota T, Miura M, Tominaga Y, Mochida A (2000) Wind tunnel tests on the nature of regional wind flow in the 270 m square residential area, using the real model: effects of arrangement and structural patterns of buildings on the nature of regional wind flow Part 1. J Archit Plann Environ Eng 529:109–116 (in Japanese)

Kubota T, Miura M, Tominaga Y, Mochida A (2002) Standards of gross buildings coverage ratio in major cities for the planning of residential area in consideration of wind flow: effects of arrangement and structural patterns of buildings on the nature of regional wind flow Part2. J Archit Plann 556:107–114 (in Japanese)

Melbourne WH (1978) Criteria for environmental wind conditions. J Wind Eng Ind Aerodyn 3:241–249. doi:10.1016/0167-6105(78), 90013-2

Ministry of the Environment, Japan (2002) Investigation report of the environmental influence by the heat island phenomenon. http://www.env.go.jp/air/report/h15–02/. Accessed Sept 19, 2010

Ministry of the Environment, Japan (2010) Climate statics. http://www.data.jma.go.jp/obd/stats/etrn/view/annually_s.php?prec_no=44&block_no=47662&year=&month=&day=&view=a2. Accessed Sept 19, 2010

Mochida A (2009) Ventilation path. Tenki 56:571–572 (in Japanese)

Murakami S, Iwasa Y, Morikawa Y (1983) Investigation of statistical characteristics of wind at ground level and criteria for assessing wind-induced discomfort: Part III Criteria for assessing wind-induced discomfort. Trans Archit Inst Jpn 325:74–84 (in Japanese)

Murakami S, Iwasa Y, Morikawa Y (1986) Study on acceptable criteria for assessing wind environment at ground level based on residents' diaries. J Wind Eng Ind Aerodyn 24:1–18. doi:10.1016/0167-6105(86), 90069-3

Murakami S, Iwamura K, Sakamoto Y, Yashiro T, Bogaki K, Sato M, Ikaga T, Endo J (2004) CASBEE; Comprehensive assessment system for building environmental efficiency. J Archit Build Sci 20:199–204 (in Japanese)

Narita K (2006) Ventilation path and urban climate. Wind Eng JAWE 1:109–114 (in Japanese)

Ng E (2009) Air ventilation assessment for high density city—an experiment from Hong Kong. In: The seventh international conference on urban climate, Yokohama, Japan

Ng E, Ren C (2009) Urban climatic mapping in Hong Kong. In: Second international conference on countermeasures to urban heat island (SICCUHI), Berkeley, California

Oguro M, Murakami S, Morikawa Y, Mochida A, Ashie Y, Ooka R, Yoshida S, Ono K (2002) A study on design methodology for outdoor thermal environment based on CFD analyses: analyses of the effect of wind from rivers on outdoor thermal environment. J Archit Build Sci 16:185–190 (in Japanese)

Oguro M, Morikawa Y, Murakami S, Matsunawa K, Mochida A, Hayashi H (2008) Development of a wind environment database in Tokyo for a comprehensive assessment system for heat island relaxation measures. J Wind Eng Ind Aerodyn 96:1591–1602

Planning Department, The Government of the Hong Kong Special Administrative Region, http://www.pland.gov.hk/pland_en/info_serv/ava_register/index.html. Accessed Jul 16, 2010

The Building Center of Japan (2009) The Building Standard Law of Japan May 2009 on CD-ROM. The Building Center of Japan, Tokyo

USGBC (2009) LEED reference guide for green building design and construction, 2009 edn. U.S. Green Building Council, Washington, DC

Yates A, Baldwin R, Howard N, Rao S (1998) BREEAM 98 for offices: an environmental assessment method for office buildings (BRE reports). IHS BRE Press, Bracknell

Yoshie R (2008) Influence of forms of cities on urban ventilation: a review. Wind Eng 33:313–320 (in Japanese)

Yoshie R, Tanaka H, Shirasawa T (2008) Experimental study on air ventilation in a built-up area with closely-packed high-rise buildings. In: Proceedings of the 4th international conference on advances in wind and structures (AWAS'08), Jeju, Korea

Chapter 7
New Criteria for Assessing the Local Wind Environment at the Pedestrian Level and the Applications

Shinsuke Kato, Zhen Bu, and Mahmoud Farghaly Bady Mohammed

Abstract The wind environment is a stochastic phenomenon. To deal with this stochastic feature, this chapter introduces the use of exceedance probability analysis to evaluate the properties of the wind. Two representative indices were adopted: the local purging flow rate (LPFR) and the space-averaged kinetic energy of the wind (KE). The former was used to express the ability of airflow to exhaust contaminants generated within a given space, and the latter was used to represent the ability of airflow to cool the human body. The local air change rate (LACR), defined as the local purging flow rate per volume of void space, was used as the ventilation index for application. The exceedance probability (EP) was calculated based on the wind speed distributions described by the Weibull function for 16 azimuths to indicate the exceedance probability of a given air change rate. The exceedance probability of the kinetic energy of the wind in the void space was also evaluated in the same manner as the local air change rate. Acceptable values of the exceedance probability for both the local air change rate and the space-averaged kinetic energy of the wind in the void space were suggested based on acceptable indoor environmental indices.

Keywords Outdoor air quality • Void • Exceedance probability analysis • Air change rate • Kinetic energy

S. Kato (✉)
Institute of Industrial Science, The University of Tokyo, 4-6-1,
Komaba, Meguro-ku, Tokyo 1538505, Japan
e-mail: kato@iis.u-tokyo.ac.jp

Z. Bu
Mott MacDonald (Shanghai, China), Unit 2601, 398 Caoxi Bei Road,
Xuhui District, Shanghai 200030, China
e-mail: buzhen@gmail.com

M.F.B. Mohammed
Faculty of Engineering, Assiut University, Assiut 271516, Egypt
e-mail: mbady@aun.edu.eg

S. Kato and K. Hiyama (eds.), *Ventilating Cities: Air-flow Criteria
for Healthy and Comfortable Urban Living*, Springer Geography,
DOI 10.1007/978-94-007-2771-7_7, © Springer Science+Business Media B.V. 2012

7.1   Introduction

7.1.1   Surface Boundary Layer

Forming the bottom level of the planetary boundary layer surrounding the Earth, the surface boundary layer extending from ground level up to several times the height of surrounding buildings is a region of particular concern for human habitation and is the area that we are particularly concerned with in this study. Because of the strong friction produced by the ground when it is covered with buildings, the wind velocity decreases rapidly from the uppermost regions down to the ground level and becomes highly turbulent because of the roughness and the topographical changes that buildings produce. The wind characteristics within this region depend largely upon ground roughness features, such as the arrangement, height, and shape of buildings, all of which contribute to the development of corresponding internal sublayers. Numerous horizontally extended buildings in urban areas create their own wind environments, i.e., their own internal sublayers.

The surface boundary layer extending from the ground level up to several times the height of the surrounding buildings is of particular concern. As the arrangement, density, shape, and orientation of buildings within an urban area can hinder the wind stream by acting as windbreaks, problems with stagnant wind conditions caused by such buildings must be examined in terms of both indoor and outdoor environments. In addition, people's activities inevitably generate heat and contaminants, among other products. These waste contaminants are often discharged indoors initially, within buildings, and then exhausted to the outside by means of the ventilation system connecting the inside environment to the outside environment. The urban wind flow between densely packed buildings may be weakened and less able to dilute or transport these waste contaminants away, resulting in higher concentrations of contaminants both inside and outside the buildings.

7.1.2   Using the Frequency of Weak or Strong Winds to Evaluate Wind Environments

People in a built-up urban area may not recognize the difference in the wind environment as compared with suburban areas with low densities of buildings. However, densely packed urban buildings cause a significant change in the wind environment. These buildings hinder the wind stream by acting as windbreaks. They weaken the average wind velocity overall and strengthen turbulent wind motion. Even though people may not recognize the slight change in the average velocity, they are sure to notice the frequency of low-velocity wind days in hot and humid seasons when the wind does not assist in reducing the heat, humidity, and thermal sensation. The wind environment is likely to be greatly affected when a single high-rise building is surrounded by several relatively low-rise buildings. The tall building can induce a strong wind, and the pedestrian-level wind environment can be significantly affected within a limited radius surrounding such tall buildings. This type of strong wind feature can be evaluated in terms of the frequency of windy days instead of the average wind velocity. The wind environment is a stochastic phenomenon (Pietrzyk and Hagentoft 2008a), and some days are characterized by strong winds, while other days are characterized by weak winds. People may not recognize slight changes in the annual mean wind velocity, but they do notice the frequency of extreme high-velocity wind days or extreme low-velocity wind days. To deal with this stochastic feature, exceedance probability analysis can be used to evaluate the properties that are affected by the dynamic features of wind.

7.1.3   The Ability of the Wind to Dilute or Transport Contaminants and to Cool Human Bodies

Buildings affect pedestrians' thermal comfort, the outdoor environment, and the indoor environment. The problem of strong winds induced by a single tall building must be addressed primarily from the point of view of pedestrians, especially in terms of the outside wind environment. From the point of view of the indoor environment, strong winds do not flow inside buildings, and they therefore cause few problems. When the wind is stagnant, effects on indoor environments will be larger and more widespread. Human activities are inevitably accompanied by the generation of heat and contaminants, which are usually initially discharged within buildings and then exhausted to the outside. A weakened urban wind may be less able to dilute or transport these contaminants away, which could result in higher concentrations of contaminants both inside and outside buildings. Some of the problems relating to this weakened urban wind are discussed below.

7.1.4   Evaluating Wind Environment Using Ventilation Efficiency in a Finite Space

Wind velocity is a vector and does not directly express the ability of wind to dilute and transport contaminants. Close to the ground, the wind velocity can change rapidly in between buildings, and it is difficult to find a single point that can adequately represent the characteristic wind environment within that vicinity. However, because the wind environment is scalar, we can use kinetic energy to represent the wind magnitude instead of its velocity. The wind environment can then be evaluated not just at a specific point but also within a finite space where people can feel its effects. To adequately express the ability of wind to dilute and transport contaminants within the finite space between buildings, this study introduces the use of a ventilation efficiency index, which can be defined within a given space, and the averaged kinetic energy, which can be used instead of point values for velocity vectors.

7.1.5   Determining an Acceptable Wind Environment Using Stochastic Evaluation

The wind environment will differ from city to city. In some cities, the wind will be relatively strong, while in other cities, it will be relatively weak. In a city where there is a relatively strong wind throughout the year, buildings can be crowded together, and the density of the buildings can be increased. In contrast, in a city where there is a relatively weak wind throughout the year, the density of the buildings should be limited to some extent to prevent the creation of a stagnant wind environment. A recommendation regarding minimum requirements for the wind environment is introduced later in this study. We believe that the urban building density should be controlled in keeping with this recommendation.

7.2   Indoor and Outdoor Air Quality in Confined Spaces

7.2.1   The Ten Times Rule for Concentration

Indoor air is mixed with outdoor air in the process of ventilation. The indoor air quality depends on the outdoor air quality, indoor pollutant sources, and the air-cleaning ability of the HVAC system. People spend more than 90% of their time indoors, and the indoor environment should be controlled so that it does not adversely affect human health. From an indoor environmental control engineering perspective, the outdoor air quality is usually expected to be ten times cleaner than the indoor air. If people want to keep indoor pollutant concentrations within the recommended guidelines using ventilation and/or other possible measures, then the outdoor pollutant concentration should be less than one tenth of the recommended value so that adequate control can be attained by means of ventilation alone. However, the indoor air quality may also vary to some extent due to indoor conditions. If the variation range of the outdoor air quality is ten times smaller and cleaner than that of the indoor air, then ventilation can be an effective measure to adequately control the indoor air quality, even in the presence of an indoor pollutant source.

Figure 7.1 illustrates the concept of the ten times rule. Region A contains a source of pollution and is ventilated into region B. Region A is enclosed within region B. The pollutant concentration in region A is required to be less than

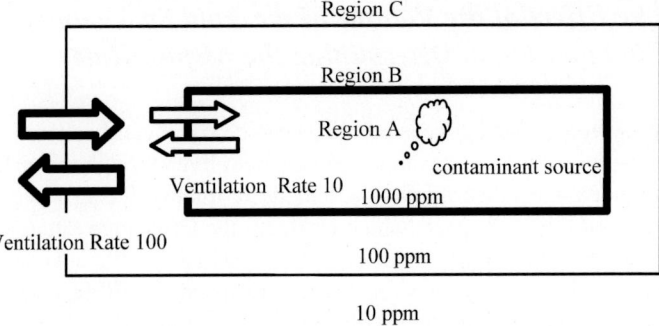

Fig. 7.1 Ten times rule for one digit accuracy

1,000 ppm. Region B is ventilated by way of both the inner region A and the outer region C. If the desired pollutant concentration in region A is to be controlled with an accuracy of 10%, then the concentration in region B should be less than 100 ppm, 10 times lower than the targeted concentration in region A, which contains the pollutant source. If the source intensity is high in region A, then the ventilation rate should be correspondingly high. The pollutant concentration in region A can be controlled with 10% accuracy if the pollutant concentration in region B is less than one tenth as high as that in region A. The same argument can be applied to the relationship between region B and region C. If region A is a building, then region B can be considered to be the local vicinity outside of the building, and region C can be considered to be the urban wind environment.

7.2.2 The Ten Times Rule for Airflow Rate

Following the same line of reasoning as that mentioned above, we may conclude that the ventilation rate for the outside air region where the building ventilation air intake is located should be ten times larger than the indoor ventilation rate when there is no outside pollutant source. If the ventilation air is being taken from the void space (Ishida et al. 2005) between buildings with a building ventilation rate of, for example, 30,000 m³/h, then this outside void space should have an airflow rate of 300,000 m³/h (with no outdoor pollutant source and no pollutant migration from other void spaces). However, the relationship between a building and the void space may not always be so clear-cut. A building can be associated with more than one void space, and one void space can be shared with more than one building. The void space can also be larger or smaller than that described above. The only critical requirement is that the outside airflow rate should be at least ten times larger than that of the indoor environment. Although the void space is relatively small compared to the scale of the building, it requires a large airflow rate to sustain the building's ventilation requirements.

7.2.3   The Effect of the Volume of the Void Space and the Air Change Rate in Determining the Airflow Rate

The size of the void space between buildings can be significantly altered by varying the arrangement of the buildings in an urban block. Even though the airflow rate is an essential index when determining the outside air quality, the airflow rate divided by the volume of the void space, which produces the air change rate (expressed as the airflow rate per unit volume), can also be a good index of the outside air quality. The air change rate expresses how often the void space air will be exchanged with the outside void space air, per unit volume, and the larger the value obtained, the better the space is ventilated and the cleaner the air. It is, therefore, useful to determine the minimum value of the air change rate that is needed for the void space between buildings to maintain a good air quality.

7.2.4   Guideline for Determining the Minimum Air Change Rate Required for a Set Volume of Void Space

If the minimum value of the air change rate for the void space can be determined and used as a guideline for the outdoor air quality in an area of densely packed buildings, and the required airflow rate can be calculated for a given building, then the void space required for that building can be determined. The larger the airflow rate required for the building, the larger the void space must be. Recall that when using the ten times rule for the airflow rate, the airflow rate for the void space must be ten times larger than the building intake airflow rate. As the size and shape of the void space is determined by the arrangement of the associated buildings, the calculation of the required void space can also be used to determine the optimum arrangement of the associated buildings. The determination of the minimum value of the air change rate required for the void space between buildings will, therefore, significantly affect the required void space for those buildings, which will, in turn, also influence the arrangement of the buildings. If the building ventilation rate is 30,000 m^3/h, as used in the previous example, then the outside void space should have an airflow rate of 300,000 m^3/h. When the minimum air change rate for the void space is 60/h, the required volume of the void space for the building is 5,000 m^3. When the building height is 10 m, the required area of the void space is 500 m^2. This void space is for the exclusive use of the building with an air intake of 30,000 m^3/h and cannot be shared with any other buildings.

   It is natural that the air change rate of the void space should be determined from the pedestrian's perspective and that the outdoor air quality and the presence of outdoor pollutant sources should be considered. If the minimum value of the void space is determined and used as a guideline for the outdoor air quality, then the level of pollutant generation should be regulated so that it does not exceed the pollutant concentration guidelines when adopting the product of the volume of the void space and the air change rate as the airflow rate.

Therefore, determining the minimum value of the air change rate for an urban void space surrounded by buildings is important for two reasons: regulating the generation of pollutants within the void space and regulating the required volume of the void space.

7.2.5   The Cooling Effect of Wind on the Human Body

Wind that is at a lower temperature than the human skin can cool the body and provide a useful means of regulating thermal comfort in the summer. Many people also claim that the utilization of wind-induced cross ventilation can reduce the cooling energy needed for buildings. It is apparent that wind-induced ventilation utilizes both kinetic and potential (pressure) wind energy. In the field of wind-induced ventilation engineering, the static pressure induced by wind at a wall surface is an important factor that is used for evaluating wind-induced ventilation. The wind static pressure is transformed from the kinetic energy of the free wind stream wherever the wind velocity falls to zero on a wall surface so that the kinetic energy also becomes zero. The capacity for wind-induced cross ventilation can therefore be represented using the kinetic energy of the void space surrounded by buildings.

7.2.6   The Ten Times Rule for Wind-Induced Cross Ventilation

The rising stream of air around a person's body, created naturally by human metabolism, is estimated to have a velocity of less than 0.3 m/s. In other words, any wind velocity greater than 0.3 m/s will cool down the human body more efficiently than will the natural convection generated by body heat. We can assume that a wind kinetic energy greater than 0.05 m^2/s^2 (approximately 0.3 m/s) will, therefore, be a useful means of cooling down the human body in any given space. If we conservatively assume that the minimum requirement for outdoor wind energy is ten times larger than the requirement indoors, allowing for the various possible conditions of wind-induced ventilation, then it follows that the indoor cross ventilation utilizes one tenth of the outdoor wind kinetic energy that is available. Therefore, the suggested minimum requirement for the efficient utilization of cross ventilation would be a value of more than 0.5 m^2/s^2 (approximately 1 m/s) for the wind energy outside of the building. As with the analogous discussion of indoor contamination control by means of ventilation, described above, we may expect that efficient indoor cross ventilation could occur wherever the outside wind kinetic energy is ten times larger than that inside.

Fig. 7.2 Void spaces in a built-up urban area

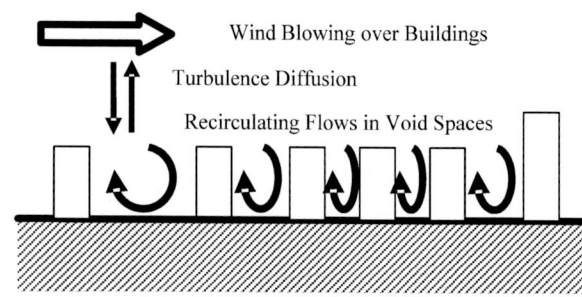

7.2.7 Voids Surrounding Buildings

In void spaces in built-up urban areas, there may be many air pollutant emissions from both the outside and inside of buildings. If these pollutants are emitted from high exhaust stacks in the upper part of the atmospheric boundary layer, then the theory of diffusion in boundary layers can be applied, and the distribution of pollutant concentrations can be predicted with a reasonable degree of reliability, even in built-up urban areas. However, pollutant diffusion within urban blocks, i.e., the diffusion of pollutants that are emitted in the lower tiers of the void space, is greatly affected by the presence of buildings and is difficult to predict (Sini et al. 1996). Figure 7.2 illustrates the characteristics of the void spaces in built-up urban areas. It is difficult to evaluate the air change rate and the space-averaged kinetic energy of wind within such a void space. Complicated wind tunnel experiments involving a scale model of an urban location or 3D CFD (three-dimensional computational fluid dynamics) are needed.

From the viewpoint of thermal comfort inside a building utilizing cross ventilation, the kinetic energy of the wind in the void spaces should be greater than 0.5 m^2/s^2 (approximately 1 m/s). If the kinetic energy is less than this, then we cannot expect a general cooling effect to be achieved by cross ventilation alone. From the viewpoint of contaminant control inside buildings, the next question is what airflow rate per unit volume, or air change rate, is needed within the void spaces. The answer should not be derived from normal environmental conditions but rather from abnormal conditions, for example, when hazardous materials are released inside a building by accident and rapid ventilation is adopted to decrease the concentration by opening windows (i.e., emergency cross ventilation is attempted). A nominal time constant of 6 ACH (the air change rate per hour) equates to a period of 10 min, and we can expect twice the nominal time constant, 20 min, to be required for the complete exchange of room air with outdoor air under complete room air mixing conditions. Therefore, 10–20 min should be the minimum/maximum time required to purge a one-shot release of hazardous materials and should, ideally, correspond to the standard emergency response time. Within this amount of time, fire fighters and ambulance crews will hopefully be able to reach the affected room and deal with the accident. To ensure that the 6 ACH rate is achieved during emergency cross ventilation in a room with open windows, the airflow rate per unit volume, or the air change rate, should be more than 60 ACH in void spaces with the same volume as the room itself. If the room faces a smaller void space, then a larger air change rate will be required.

A nominal time constant of 60 ACH equates to a period of 1 min, and we can expect 2 min to be needed for the complete exchange of void space air with air from outside the void. Most people should be able to hold their breath for a minute so as not to inhale any pollutant that is accidentally discharged outside. In the case of a one-shot release of a hazardous material inside a building, the inhabitants can hopefully escape to safety outside. It is assumed that the outside will be safer than the inside, even when we have such a release outdoors. This means that the area outside the buildings should be 10 times safer than that inside the buildings.

7.2.8 Contaminants Released in Void Spaces

Contaminants may be generated outside the buildings as well as inside. When we consider the air change rates that are needed in void spaces to purge any contaminants present, we should also consider the nature of the contaminant generated and its ability to transit the void space. There are two types of contaminant generation and two types of contaminant purging. One type of contamination involves the generation of the contaminant directly in the void space itself, while the other involves a contaminant that is transported there from an upstream void. The differences between these two types of contaminant generation should not be overlooked, and each should be estimated and addressed separately. We should aim to decrease the degree of contaminant generation in the void of interest and also decrease the amount of contaminant migration from upstream sites. These are two separate tasks, and the contaminant concentration within the void is determined by these two categories of contaminant generation; the required purging flow rate of the void is determined accordingly.

The amount of contaminant that is generated on-site or that migrates in can be purged both to the upper tiers of the void space and to the next downstream void. The purging of the former should occur at a higher rate than the latter, and, in urban areas, the pollution generated should be purged to the upper tiers if possible and not to the downstream void. In this context, the characteristics of turbulent diffusion to the upper tiers are especially important and must be estimated exactly for void spaces in built-up urban areas. 3D CFD can be used to execute the complicated tasks required for this type of calculation.

7.3 New Criteria for Assessing the Local Wind Environment at the Pedestrian Level Based on Exceedance Probability Analysis

7.3.1 Exceedance Probability Analysis

The wind environment is a stochastic phenomenon. To deal with this stochastic feature, the present study introduces the concept of exceedance probability analysis to evaluate the extreme properties of the wind instead of the mean values. People may not recognize the daily change in the average wind velocity, but they are sure to notice the frequency of stagnant wind days with hot thermal conditions and/or a highly polluted atmosphere. They will be able to evaluate the wind environment in terms of the frequency of low-velocity wind days when the wind did not noticeably reduce the heat and humidity and/or the amount of air pollution. The wind environment can be evaluated mathematically using a probability density function rather than just exact values, such as the mean or variance. It is important to determine how often stagnant wind days occur. We can use the probability density function of the instantaneous properties evaluated with a three-second response time (0.33 Hz time resolution), but hourly mean values (0.3×10^{-3} Hz time resolution) are usually utilized. The former is used for evaluating instantaneous strong wind events, which can greatly affect pedestrian safety. In practice, as it is difficult to predict instantaneous wind features with a wind tunnel experiment or CFD, hourly mean values are used, and a gust factor is introduced to estimate instantaneous features from the mean values. The latter is used for evaluating stagnant wind conditions, which are important when assessing the purging of contamination and/ or controlling the thermal environment to maintain comfort levels. In the wind engineering field, hourly mean values are usually used for evaluations of both strong winds and stagnant wind.

It is generally accepted that the anemometers that are generally used at weather stations cannot measure low wind velocities of less than 1 m/s. Above a wind speed of 1 m/s, these anemometers have a measurement resolution of 0.1 m/s, but below 1 m/s, they are only able to show that the wind velocity is less than this value (with a measurement resolution of 1 m/s).

In past studies, a probability analysis has been applied in the investigation of the pedestrian-level wind environment around buildings; they have generally focused on safety issues. Penwarden and Wise (1975), Hunt et al. (1976), Murakami et al. (1986), and Ohba et al. (1988) studied assessment criteria from the viewpoint of the acceptable frequency of the occurrence of wind speed. For instance, Murakami et al. (1986) used an exceedance probability (EP) analysis to construct criteria to determine the acceptable frequency of strong winds in a built-up area in Tokyo, and the exceedance frequencies of the daily maximum wind speed and the daily maximum gust wind speed were analyzed with the speed described by the Weibull distribution. All of the above studies applied local scalar velocity as the main index in their probability analyses to investigate the effects of strong winds at the pedestrian level. Meanwhile, other studies applied several domain-averaged indices, such as the purging flow rate (PFR) (Huang et al. 2006) and SET* (Ishida et al. 2005), to evaluate the overall ventilation performances of the investigated domain. Nevertheless, because these studies only used deterministic approaches rather than a probabilistic analysis, they only looked at certain wind velocities or wind directions and could not realistically describe the random outdoor ventilation phenomena that occur in nature. Recently, Pietrzyk and Hagentoft (2008b) presented a probabilistic model by using the air change rate as an index based on a theoretical calculation to study the problem of air infiltration in low-rise buildings. Few reports based on a similar probability analysis with the investigation of outdoor wind-driven natural ventilation performance exist in the literature. As an extension from this velocity index, the LACR and KE are applied as new indices, as are two corresponding criteria; the LACR-EP and KE-EP are proposed in this study to investigate the overall characteristics of the local wind environment of void spaces.

7.3.2 Velocity-Based Exceedance Probability (V-EP)

The V-EP criterion is employed to investigate the impacts of strong winds on pedestrian comfort and to assess the acceptable levels to be used in urban planning and design. Because statistical data on the wind speeds and directions at ground level is often unavailable for regions of interest, a local meteorological station is usually selected as the reference point. Generally, wind velocity ratios are required to calculate the V-EP for 16 azimuths to transform the meteorological data from the reference site to local site at ground level. As indicated in Eq. 7.1, the velocity ratio $R(a_n)$ is defined as the ratio between the scalar velocity at ground level V_g and the velocity at reference height $V(a_n)$ for azimuth a_n.

$$R(a_n) = \frac{V_g}{V(a_n)} \tag{7.1}$$

The statistical data on the wind speed $V(a_n)$ for 16 azimuths are assumed to follow a two-parameter Weibull distribution for each azimuth, given as

$$P(>V(a_n)) = A(a_n) \times \exp\left\{ -\left(\frac{V(a_n)}{C(a_n)}\right)^{K(a_n)} \right\} \tag{7.2}$$

where $P(>V(a_n))$ is the probability of exceeding a given wind speed $V(a_n)$ at a reference height for azimuth a_n, $A(a_n)$ is the relative frequency of occurrence, and $K(a_n)$ and $C(a_n)$ are two parameters of the Weibull distribution.

Based on similarity theory of the linear relationship between the wind speed at ground level and that at the reference level, the velocity ratio for each azimuth can be assumed to be a constant value in most cases and can be determined by direct measurement, wind tunnel experimentation, or CFD simulation. As also indicated by Pietrzyk and Hagentoft (2008b), the shape parameter of the Weibull distribution for the site is the same as for the meteorological station. Consequently, from Eqs. 7.1 and 7.2, the EP of the wind velocity at the ground level can be obtained for each azimuth direction. By summing the EP for all of the 16 azimuths, the overall V-EP at the ground level is expressed by the following equation:

$$P(V > V_g) = \sum_{n=0}^{15} A(a_n) \times \exp\left\{ -\left(\frac{V_g}{R(a_n) \times C(a_n)} \right)^{K(a_n)} \right\} \tag{7.3}$$

Equation 7.3 represents the overall probability of exceeding a given scalar velocity V_g at the ground level throughout all the azimuth directions, which is applied as a criterion to assess critical value of the wind speed at the ground level to ensure that the wind environment of the considered site is acceptable for outdoor human activities. Equation 7.3 also indicates that the V-EP criterion uses the site velocity V_g as the index, which means that it is a point-based criterion that is suitable for assessing a point-based physical index, such as scalar velocity.

Nevertheless, due to the randomness and the unevenness that is associated with outdoor airflow in most cases, the entire considered domain, rather than only some key points, should be taken into consideration when assessing the local wind environment with regard to ventilation efficiency or thermal comfort. From this aspect, the spatial average value of the target domain can be used to represent the characteristics of the local wind environment for convenience, whereas point values are more suitable for the investigation of risk evaluations, distribution properties, and other purposes. Namely, the domain-based index that is able to represent the ventilation performance or thermal comfort level of the target domain should be selected and applied to construct corresponding domain-based criteria. For this purpose, two domain-based criteria, LACR-EP and KE-EP, are introduced in the following sections by using the domain-based indices LACR and KE, respectively. The newly proposed criteria have an advantage over the traditional velocity-based criterion in that they are capable of estimating the comprehensive characteristics of the wind environment at the pedestrian level.

7.3.3   Local Air Change Rate-Based Exceedance Probability (LACR-EP)

7.3.3.1   Purging Flow Rate and Local Air Change Rate

The first index introduced is the local air change rate. As shown in Eq. 7.4, the LACR is calculated by the PFR and the volume of the target domain. The PFR was originally defined as the effective airflow that is required to remove/purge the local pollutants; it was proposed by Sandberg and Sj berg (1983) and was used as an index to evaluate the ventilation performance and air quality in indoor airflow problems (Sandberg and Sj berg 1983; Peng and Davidson 1997; Ito et al. 2000). Following this concept, the PFR was also considered to be effective with respect to the assessment of the outdoor airflow problem. As defined in Eq. 7.5, the PFR only represents the effective airflow for the whole void domain from the perspective of the capacity of the wind to disperse the interior pollutants. From a more practical perspective, the LACR index represents the average ventilation efficiency of the same target domain.

$$LACR = \frac{3600 \times PFR}{vol} \tag{7.4}$$

$$PFR = \frac{q}{c} \tag{7.5}$$

where q is the spatially uniform generation rate of the pollutant source (kg/s) and c is average concentration of the entire target domain (kg/m^3).

7.3.3.2   LACR-EP Criterion

As shown in Fig. 7.4 in Sect. 7.4, for a void model with orientation β and upper wind from azimuth a_n, the wind can be regarded as approaching from direction i relative to the orientation of the street ($i = n - \beta$). In this case, the wind speed at the reference height and the air change rate of the target domain at ground level are expressed as $V_0(a_n)$ and $LACR_{0i}$, respectively. As the air change rate is proportional to the local wind velocity in most cases, the required reference wind speed $V(a_n)$ required for azimuth a_n to satisfy the specified air change rate $LACR_g$ at the ground void domain can be calculated by Eq. 7.6. Meanwhile, we define the ratio of $LACR_{0i}$ to $V_0(a_n)$ as $R_N(a_n)$, given in Eq. 7.7 using the analogy of the definition of the velocity ratio $R(a_n)$. Although $R_N(a_n)$ has no clear physical meaning, it can be easily calculated by CFD or measured by a wind tunnel.

$$V(a_n) = \frac{LACR_g}{LACR_{0i}} \times V_0(a_n) = \frac{LACR_g}{LACR_{0i}/V_0(a_n)} \tag{7.6}$$

$$R_N(a_n) = \frac{LACR_{0i}}{V_0(a_n)} \tag{7.7}$$

Substituting Eqs. 7.6 and 7.7 into Eq. 7.2 and summing for all of the 16 azimuths, the overall LACR-EP of the void domain is expressed by

$$P(>LACR_g) = \sum_{n=0}^{15} A(a_n) \times \exp\left\{-\left(\frac{LACR_g}{R_N(a_n) \times C(a_n)}\right)^{K(a_n)}\right\} \tag{7.8}$$

where $P(>\text{LACR}_g)$ represents the probability of exceeding the specified air change rate LACR_g, which also reflects the acceptable level of the local ventilation efficiency or dispersion effect. Based on the above calculations and the LACR-EP criterion, we can estimate the local ventilation performance at the ground level by using the available statistical information from the reference height, although prior to now, there has been no practical limit value proposed for assessing the outdoor wind environment as it affects ventilation efficiency.

7.3.4 Local Kinetic Energy-Based Exceedance Probability (KE-EP)

7.3.4.1 Local Kinetic Energy

Regarding the mechanism of air exchange between the void space and the outer surroundings, turbulent diffusion is the main factor dominating this process, which is actually a phenomenon of the turbulent transport of pollutants as well as the turbulent transport of momentum and energy. Thus, the wind environment of the void space may also be investigated from the perspective of energy. In this section, another index, the local kinetic energy, is also introduced in the study of the local wind environment in addition to the air change rate presented above. The local kinetic energy is given by the following equation:

$$KE = \frac{1}{vol} \iiint_{void} \left(\frac{1}{2} \times \left(\overline{U}^2 + \overline{V}^2 + \overline{W}^2 \right) + k \right) dv \tag{7.9}$$

where KE is the spatially averaged kinetic energy (m^2/s^2); $\overline{U}$, $\overline{V}$, and $\overline{W}$ are the averaged velocity components (m/s); and k is the averaged turbulent kinetic energy (m^2/s^2) of the entire void domain.

The kinetic energy represents the performance of the wind in improving the thermal comfort level from the aspect of wind intensity. As indicated in Eq. 7.9, the KE index includes both the averaged velocity components and the turbulent components, which means that the kinetic energy can evaluate the convective effects of wind more comprehensively and precisely than can other indices, such as the scale velocity. This extra precision is due to the dependence of the thermal comfort level on the average velocity as well as the wind turbulence, especially in domains where the wind turbulence dominates the flow field.

7.3.4.2 KE-EP Criterion

In a manner similar to the air change rate-based EP criterion, the exceedance probability can be expressed in terms of the kinetic energy. At first, with the wind speed $V_0(a_n)$ defined at the reference height, the corresponding kinetic energy of the target domain is expressed as KE_{0i} at the ground level. Based on a dimensional analysis, the kinetic energy is proportional to the square value of the wind velocity. Therefore, the required reference wind speed $V(a_n)$ for azimuth a_n to satisfy the specified KE_g of the void domain at the ground level can be estimated by Eq. 7.10. Similarly, the ratio of KE_{0i} to $V_0(a_n)$ is defined as $R_K(a_n)$, as given in Eq. 7.11.

$$V(a_n) = \sqrt{\frac{KE_g}{KE_{0i}}} \times V_0(a_n) = \frac{\sqrt{KE_g}}{\sqrt{KE_{0i}}/V_0(a_n)} \tag{7.10}$$

$$R_K(a_n) = \frac{\sqrt{KE_{0i}}}{V_0(a_n)} \tag{7.11}$$

By substituting the above two equations into Eq. 7.2, the overall KE-EP of the void domain for all of the 16 azimuths can be calculated as follows:

$$P\left(>KE_g\right) = \sum_{n=0}^{15} A(a_n) \times \exp\left\{-\left(\frac{\sqrt{KE_g}}{R_K(a_n) \times C(a_n)}\right)^{K(a_n)}\right\} \tag{7.12}$$

where $P(>KE_g)$ is the probability of exceeding the specified KE_g. This kinetic energy-based EP reflects the overall acceptable level of the local thermal environment.

These newly proposed criteria have an advantage over the traditional velocity-based criterion because they are capable of estimating the comprehensive characteristics of the wind environment at the pedestrian level.

7.3.5 Calculation Procedure and Application Method of New Criteria

As indicated in Eqs. 7.8 and 7.12, it is necessary to obtain the value of $R_N(a_n)$ and $R_K(a_n)$ in addition to other parameters, including $A(a_n)$, $C(a_n)$, and $K(a_n)$, to determine the LACR-EP and KE-EP. As mentioned in the former section, $R_N(a_n)$ and $R_K(a_n)$ can be determined by CFD simulation or by a wind tunnel experiment. However, due to the difficulties and complications involved in measuring the physical quantities of the target domain, CFD simulation is implemented to calculate the indices of the LACR and KE in this study.

However, as a meteorological station is often selected as the reference point, the Weibull parameters $C(a_n)$ and $K(a_n)$ and frequency of occurrence $A(a_n)$ are available from the lengthy meteorological observation record. When changing the reference height, for example, when investigating and comparing a building model in a fixed geometry for two cities, there is no need to recalculate the $R_N(a_n), R_K(a_n)$, $LACR_{0i}$, or KE_{0i} for each azimuth. We can select one city as the base city and the other as the target city. The reference height and the reference velocity of the base city are indicated by h_B and $V_B(a_n)$, respectively, with the subscript of "B." Likewise, the corresponding height and velocity of the other city are expressed as h_T and $V_T(a_n)$, respectively, with the subscript of "T." Based on the assumption that the approaching wind has a velocity profile of the power law with an exponent of α ($\alpha = 0.25$ for the city), Eqs. 7.6 and 7.10 become:

$$V_T(a_n) = V_B(a_n) \times \left(\frac{h_T}{h_B}\right)^\alpha = \frac{LACR_g}{LACR_{0i,B}/V_{0,B}(a_n)} \times \left(\frac{h_T}{h_B}\right)^\alpha$$

$$= \frac{LACR_g}{R_{N,B}(a_n)} \times \left(\frac{h_T}{h_B}\right)^\alpha \qquad (7.13)$$

$$V_T(a_n) = V_B(a_n) \times \left(\frac{h_T}{h_B}\right)^\alpha = \frac{\sqrt{KE_g}}{\sqrt{KE_{0i,B}}/V_{0,B}(a_n)} \times \left(\frac{h_T}{h_B}\right)^\alpha$$

$$= \frac{\sqrt{KE_g}}{R_{K,B}(a_n)} \times \left(\frac{h_T}{h_B}\right)^\alpha \qquad (7.14)$$

By substituting the above two equations into Eq. 7.2 and summing for all of the 16 azimuths, the overall LACR-EP and KE-EP for the target city become:

$$P(>LACR_g) = \sum_{n=0}^{15} A_T(a_n) \times \exp\left\{-\left(\frac{LACR_g}{R_{N,B}(a_n) \times C_T(a_n)} \times \left(\frac{h_T}{h_B}\right)^\alpha\right)^{K_T(a_n)}\right\} \qquad (7.15)$$

$$P(>KE_g) = \sum_{n=0}^{15} A_T(a_n) \times \exp\left\{-\left(\frac{\sqrt{KE_g}}{R_{K,B}(a_n) \times C_T(a_n)} \times \left(\frac{h_T}{h_B}\right)^\alpha\right)^{K_T(a_n)}\right\} \qquad (7.16)$$

With the application of the LACR-EP and KE-EP calculated by Eqs. 7.15 and 7.16, it is possible to establish an assessment system for the wind environment at the pedestrian level and to estimate the potential natural ventilation for different cities within a wide range.

7.3.6 Exceedance Probability of 1/7 or 6/7

There are 8,760 h in a year. An exceedance probability of 6/7 is approximately 7,500 times out of 8,760, meaning that only 1 day a week fails to reach a set value, which in this case is a given air change rate (LACR$_g$) or a given kinetic energy (KE$_g$). Conversely, an exceedance probability of 1/7 is approximately 1,250 times out of 8,760, meaning that 1 day a week attains a set value, such as a given air change rate or a given kinetic energy. As our day-to-day rhythms follow both a weekly activity pattern, 7 days a week, and a daily 24-h cycle, an exceedance probability of 6/7 makes it easy for people to recognize that there is 1 day of stagnant wind per week, i.e., it is an "unlucky day." Likewise, an exceedance probability of 1/7 will be obvious to people that there is 1 day of favorable wind per week, i.e., it is a "lucky day." We may, therefore, give special significance to events that occur with a probability of 1/7 or 6/7.

As mentioned earlier, the air change rate in void spaces should be more than 60 ACH, meaning that the exceedance probability should exceed 6/7 for the 60 ACH. The kinetic energy of the wind in void spaces should, preferably, be above 0.5 m^2/s^2 (approximately 1 m/s). Therefore, an exceedance probability of 1/7 for 0.5 m^2/s^2 will produce a favorable wind environment where wind-induced cross ventilation can be achieved at least once a week.

7.4 Application 1: Urban Ventilation in an Idealized Street Canyon

7.4.1 Void Model Description

Figure 7.3 shows the configuration of the street canyon model that is explored in this study. As depicted in this figure, three isolated street canyons are arranged in parallel along the streamwise direction (X coordinate) at a distance of 100 m apart. These

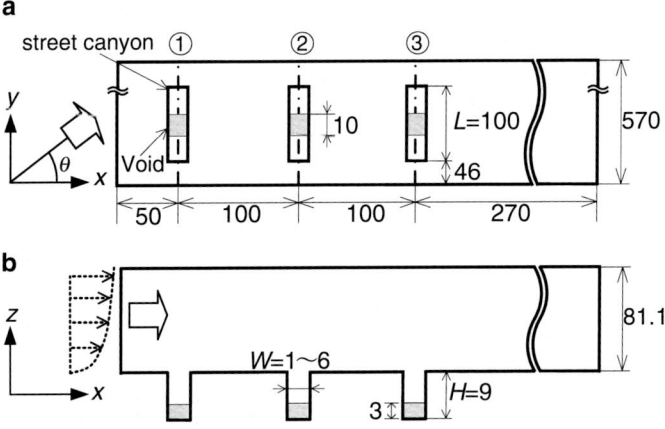

Fig. 7.3 Configuration of the model. (**a**) Horizontal plan. (**b**) Vertical plan (unit: m)

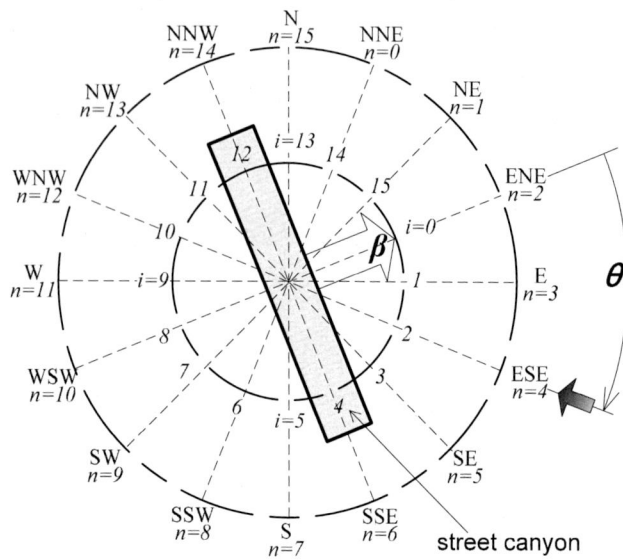

Fig. 7.4 Model definitions

three canyons are of the same dimensions, with a length (L) of 100 m, a height (H) of 9 m, and a width (W) that changes from 1 to 6 m. At the center of each canyon at the ground level, the domain W $\times$ 10 m $\times$ 3 m in the X, Y, Z direction is defined as the void space for further discussion. The arrows indicate the wind direction. In this study, this street canyon model is called a void model, representing street canyons in a typical residential complex in Tokyo, Japan. To simplify the model, all of the residential buildings were uniformly set as three stories in height, with no gap between the buildings.

7.4.2 Definitions of the Model Orientation and Incidence Angle

As shown in Fig. 7.4, the horizontal plane is divided into 16 azimuths starting from the north-northeast (NNE) in a clockwise direction, i.e., a_0 = NNE, a_1 = NE..., a_{15} = N. For the canyon model, the direction of the short side of the canyon is defined as the orientation of the model, and the angle of the orientation is represented by β. The approaching wind is grouped into 16 directions relative to the orientation of the model and is marked by i, beginning from the orientation of the canyon clockwise. θ represents the wind incidence angle.

Table 7.1   CFD analysis conditions

Turbulent model	Standard k-ε model
Differential scheme	Convection terms: MARS[a]
Inlet[b] (Murakami et al. 1998)	$V = V_B \times (h/h_B)^{1/4}$
	$k = 1.5 \times (I \times V)^2, I = 0.1$
	$\varepsilon = C_\mu \times k^{3/2}/l$
	$l = 4(C_\mu \times k)^{1/2} h_B^{1/4} h^{3/4}/V_0$
Outlet	Free outflow
Side, Top	Free slip
Ground, Wall	Generalized logarithmic law
Other	The pollutant was assumed to be passive, with a spatially uniform emission rate of 0.001 (kg/m^{3}s) within the void domain

[a]*MARS* Monotone Advection and Reconstruction Scheme, second-order scheme (CD Adapco Group 2004)
[b]Reference height $h_B = 74.5$ m; reference velocity $V_B = 1.0$ m/s

7.4.3   Index of Air Change Rate and Kinetic Energy

CFD simulations were first performed for the model shown in Fig. 7.3 to calculate the index of the local air change rate and the average kinetic energy of the void space for the 16 wind incidence directions. Moreover, two parameters, the height/width (*H/W*) aspect ratio (Hunter et al. 1992) and the incidence angle of the approaching wind θ, were investigated for the indices of LACR and KE.

7.4.3.1   Calculation Settings

Due to the symmetry of the void model in this study, the simulations were performed for only five incidence angles, 0.0°, 22.5°, 45.0°, 67.5°, and 90.0°, which are capable of representing the ventilation performance for all of the 16 wind directions. In the case that the orientation of the model is ENE, these incidence angles correspond to five azimuths, ENE, E, ESE, SE, and SSE, as shown in Fig. 7.4. The numerical simulations were performed by using the STAR-CD software, and the detailed analysis conditions are shown in Table 7.1. The location is Tokyo, Japan, and the Tokyo Meteorological Observatory is selected as the reference point, with a height of 74.5 m (Japan Association for Wind Engineering 2005).

7.4.3.2   Calculation Cases

In this study, the combination of two parameters, the *H/W* ratio and the wind incident angle θ, were studied for 30 cases. The investigated *H/W* ratio is from 1.5 to 9, with the *H* fixed at 9 m and the *W* ranging from 1 to 6 m at a pitch of 1 m; θ is between 0.0°

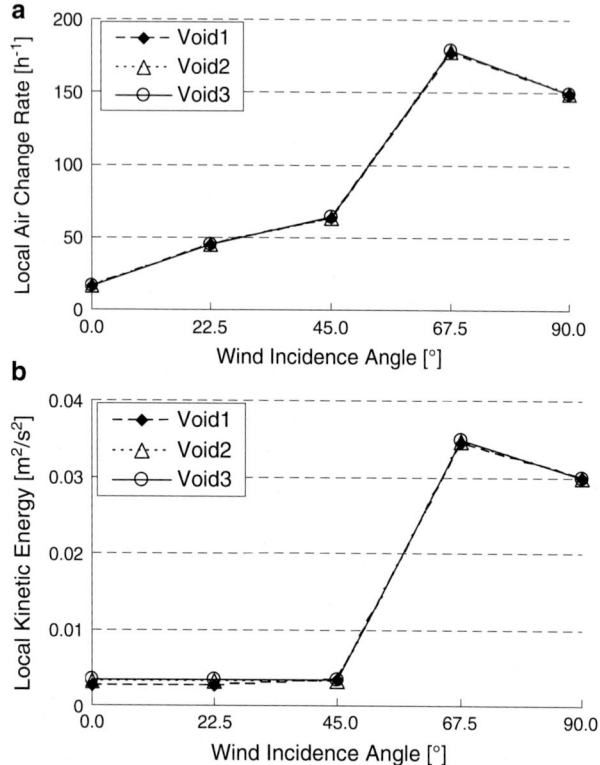

Fig. 7.5 Comparison among three void domains ($W = 4$ m). (**a**) Local air change rate. (**b**) Local kinetic energy

and 90.0° at a pitch of 22.5°. The H/W ratio has been regarded as one of the most important configuration factors affecting the microclimate within the void space. At a greater aspect ratio ($H/W > 0.7$), a stable circulatory vortex is established within the canyon for $\theta = 0.0°$, resulting in skimming flow regime (Oke 1988). All of the cases of this study are within such a regime, with the characteristics of recirculation airflow and weak dispersion of pollutants. Because H was fixed in all cases, we instead use W as the configuration parameter to investigate the influence of the H/W ratio in the following discussion.

7.4.3.3   Investigation of the Indices: LACR and KE

The three street canyons in the streamwise direction were initially compared in terms of the indices of LACR and KE. For the case of $W = 4$ m as an illustration, the calculated indices of voids 1, 2, and 3 are nearly identical, independent of the wind incidence angle (Fig. 7.5). This result indicates that there is sufficient distance between the three canyons to ensure consistency between them with regard to the

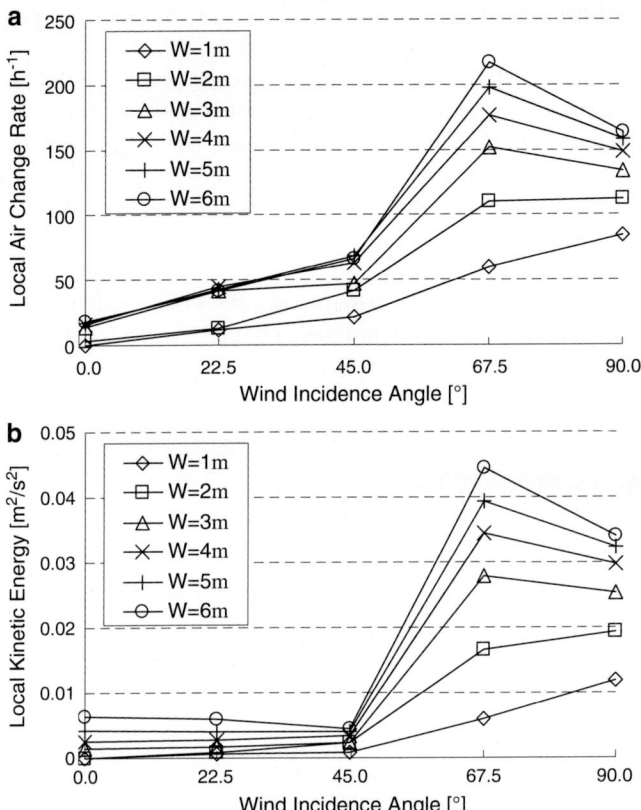

Fig. 7.6 Results of the LACR and KE. (**a**) Local air change rate. (**b**) Local kinetic energy

criteria calculated in the following section. Moreover, similar conditions were also found for the other cases. The results for void 1 are used hereafter for further discussion.

The simulation results for the local air change rate and local kinetic energy for all cases are shown in Fig. 7.6a, b, respectively. As indicated in Fig. 7.6a, the minimal value occurs for all considered widths when the wind direction is perpendicular to the street direction ($\theta = 0.0°$). Namely, the wind incidence angle of $0.0°$ gives the worst ventilation performance. As the wind incidence angle increases, the air change rate shows a tendency to increase, and it increases sharply at the incidence angle of $45.0°$. For $W > 2$ m, the peak value is reached when $\theta = 67.5°$; for $W \leq 2$ m, it is reached when $\theta = 90.0°$. For the cases with a wider width, when $\theta = 67.5°$, strong spiral flows could be induced from the upper part of the canyon, improving the ventilation performance of void domain at ground level. Nevertheless, this phenomenon becomes weak in a canyon with a narrow width, and the maximum value is obtained when the wind blows along the street. As for the influence of W, an increase in W can also help to improve the ventilation performance at the ground level, especially for the W ranging from 1 to 3 m. However, this improvement decreases for $W > 4$ m.

The results for the local kinetic energy, shown in Fig. 7.6b, show similar distributions for the most part. An exception is that there is almost no change in the kinetic energy index for a wind incidence angle θ that ranges from 0.0° to 45.0°. This result could be explained by the fact that the strong vortex circulation at $\theta = 0.0°$ permits a small increase in θ to assist in the dispersion of pollutants through the lateral faces of the void domain, whereas there is no change for the energy index KE.

In general, both environmental parameters, the wind incidence angle and the aspect ratio parameter, play important roles in improving the local microclimate of the void domain at the pedestrian level.

7.4.4 Exceedance Probability Analysis

Because the reference velocity at the reference height was fixed at 1.0 m/s for convenience in the above CFD simulation, the calculated index of the LACR and KE are equal to the R_N and R_K, respectively, based on Eqs. 7.7 and 7.11. By substituting these indices into Eqs. 7.8 and 7.12, we can obtain the probabilities of exceeding the corresponding given index. In the present section, as an illustration, the EP values were calculated for the LACR from 0 to 400 h^{-1} and for the KE from 0 to 0.5 m^2/s^2. These values were applied to investigate the influence of orientation β, which is a geometrical factor, and the H/W aspect ratio, which is a configuration factor, on the local wind environment of the void space at the pedestrian level.

7.4.4.1 Statistical Parameters

Figure 7.7 is the wind rose of Tokyo, representing the frequency of the occurrence of each wind direction $A(a_n)$ based on hourly measured statistical data for 10 years (from 1995 to 2004), recorded at the Tokyo Meteorological Observatory (Japan Association for Wind Engineering 2005). As indicated from this figure, the prevailing wind direction in Tokyo is north-northwest (NNW), occupying 20.6% of the total occurrence. Due to the symmetry of the void model, wind from one specific direction has the same effect of "cleaning" or "cooling" as that from the opposite direction. Therefore, the frequency of the occurrences from two opposite directions are coupled together to investigate the wind performance of one direction, referred to hereafter as the "wind-direction group." Therefore, the three prevailing wind-direction groups in Tokyo are SSE-NNW (22.9%), S-N (18.6%), and NE-SW (17.7%), while the nonprevailing wind-direction groups are ESE-WNW (5.6%), E-W (6.1%), and SE-NW (7.7%). The Weibull parameters of $K(a_n)$ and $C(a_n)$ are listed in Table 7.2.

Fig. 7.7 Wind rose of Tokyo

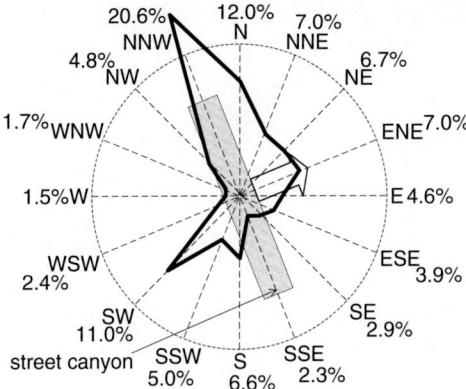

Table 7.2 Weibull parameters of Tokyo

Parameters	NNE	NE	ENE	E	ESE	SE	SSE	S
$C(a_n)$	3.34	3.52	3.6	3.28	3.02	2.64	2.48	3.67
$K(a_n)$	2.61	2.53	2.55	2.50	2.46	2.71	2.92	2.91
Parameters	SSW	SW	WSW	W	WNW	NW	NNW	N
$C(a_n)$	3.85	4.39	2.39	1.92	2.48	3.82	4.06	3.58
$K(a_n)$	2.50	2.37	2.35	2.23	1.88	1.91	2.51	2.57

7.4.4.2 Investigation of Model Orientation by Exceedance Probability Analysis

Figures 7.8a, b show the calculated distributions of the LACR-EP and KE-EP for eight orientations, NNE, NE, ENE, E, ESE, SE, SSE, and S for the void model with a width of 4 m. The calculation of the other eight orientations can be omitted due to the symmetry of the void model.

As indicated in Fig. 7.8a, the curves of the air change rate-based exceedance probabilities have quite large differences according to the orientation β. When the exceedance probability is high, the differences between the orientations will be larger. In general, the EP has the highest value when the orientation of the model is east (E), while the lowest distribution is observed at north-northeast (NNE). Additionally, as indicated in Fig. 7.6a, the ventilation performance in the void space is greatly influenced by the wind incidence angle at 67.5° and 90.0°. Consequently, for the model facing E, the dominant wind directions are SSE, S, SSW, NNW, N and NNE. These directions are to a great extent coincident with the three prevailing wind groups in Tokyo, SSE-NNW, S-N, NE-SW, and this result explains why orientation E has the highest ventilation potential. Similarly, the worst ventilation performance is found for canyons with the orientation NNE, where the wind incidence angles of 67.5° and 90.0° are coincident with the nonprevailing wind groups ESE-WNW, E-W, and SE-NW.

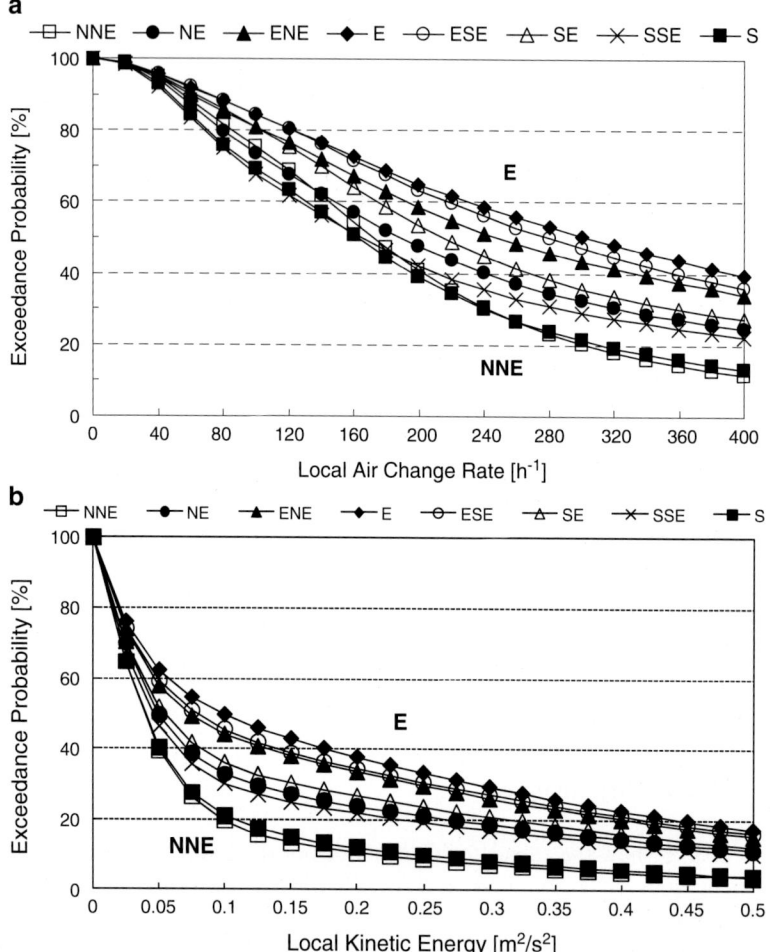

Fig. 7.8 Distributions of EP according to model orientation ($W = 4$ m). (**a**) Local air change rate-based EP. (**b**) Local kinetic energy-based EP

Similarly, the maximum value and the minimum value are also found at orientations E and NNE for the local kinetic energy-based EP criterion.

The local air change rate in the void space outside a building should be 60 ACH to ensure 6 ACH emergency cross ventilation in a room with the windows open. Figure 7.8a indicates that 60 ACH can be attained with 90% probability for a street canyon orientation of east (E) and with 83% for a canyon orientation of north-northeast (NNE) when $W = 4$ m and $H = 9$ m.

We have assumed that a wind kinetic energy greater than 0.05 m^2/s^2 (approximately 0.3 m/s) will also be useful for cooling the human body by means of wind-

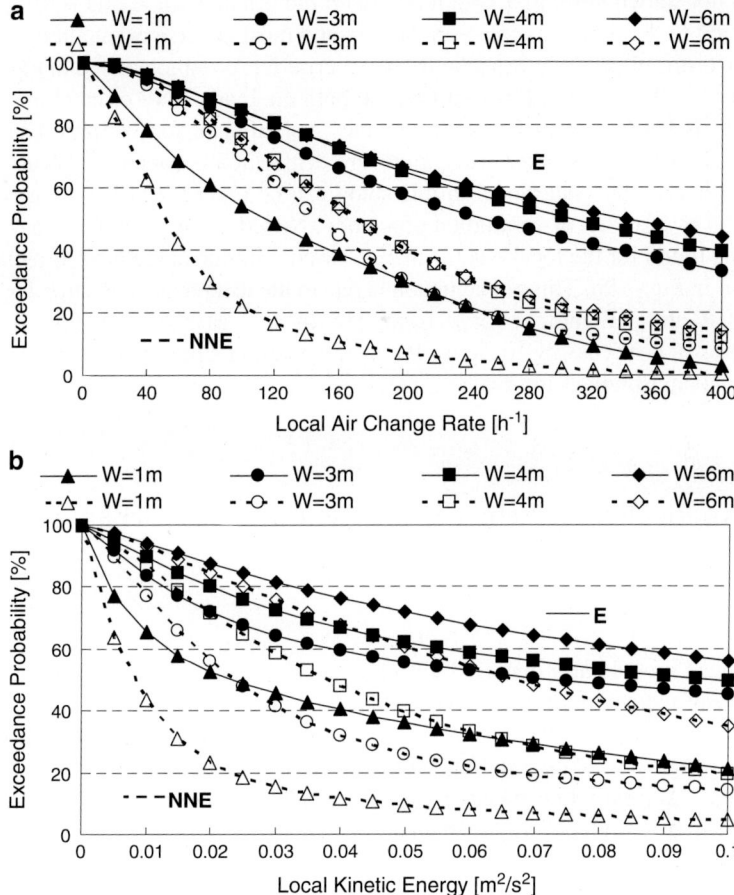

Fig. 7.9 Distributions of EP according to the model width (orientation: E, NNE). (**a**) Local air change rate-based EP. (**b**) Local kinetic energy-based EP

induced cross ventilation and that for this to occur, a minimum wind energy of 0.5 m^2/s^2 (approximately 1 m/s) is required outside of the building. Figure 7.8b indicates that such conditions can only be obtained with approximately 15% probability for most of the street canyon orientations in Tokyo when $W = 4$ m and $H = 9$ m.

7.4.4.3 Investigation of the Model Width by Exceedance Probability Analysis

Figures 7.9a, b show the calculated distributions of the criteria LACR-EP and KE-EP, respectively, with model widths ranging from 1 to 6 m (curves of 2 and

5 m are not shown here) in a void model with the orientations E and NNE, which correspond to the most favorable orientation and most adverse orientation, respectively, in terms of the local wind environment criteria. As demonstrated in Fig. 7.9a, the change in the LACR-EP is significant at both the E and NNE orientations when the width is below 3 m, whereas it becomes insignificant for W from 4 to 6 m. Because the EP for the other orientations are distributed between the E curve and the NNE curve, the void model in this study can be regarded as having a critical value of 4 m in terms of ventilation efficiency. Nevertheless, such a critical value cannot be found for the local kinetic energy-based criterion at the NNE orientation, as shown in Fig. 7.9b. This phenomenon is due to the discrepancy existing between the LACR and KE indices for θ from 0.0° to 45°, as shown in Fig. 7.6a, b. Therefore, it is necessary to apply the corresponding criteria in evaluating the local wind environment for different purposes.

7.5   Application 2: Urban Ventilation in Dense Urban Areas

7.5.1   Model Description

Four typical models of a densely built-up area are illustrated in Fig. 7.10, and the geometry of the central part of model (I) is shown in Fig. 7.11. Similar to application 1, the definitions of the dense urban area model in terms of the orientation and incidence angles are shown in Fig. 7.12.

In Fig. 7.10, the white blocks represent detached houses, while the gray area represents roads and voids between adjacent buildings. The study domain (in all of these models) is the pedestrian volume located along a street surrounded by type A and type B blocks (shown in Fig. 7.11); the blocks are surrounded by a fence of 1.5 m height. The study domain has the same dimensions as the street (40 m × 4 m) and extends to a height of 5.5 m.

Models (I), (II), and (III) have the same building arrangements and dimensions, but they differ in the geometry of the central part that surrounds the street (marked with black-dashed boxes). In model (I), the central part consists of eight type A and two type B blocks. Narrow gaps with a width of 1 m exist between the type A buildings and between the type B buildings. In model (II), the central part forms a solid U shape. In model (III), the street buildings are the same as those of model (I), while the outer blocks that surround them form a solid U shape. In model (IV), the symmetry is interrupted by introducing other blocks of different dimensions in addition to shifting the upper and the lower two thirds of the array along the x-axis.

Many factors were considered during the design of these four building patterns. First, these models suit the nature of existing small-lot residential areas in Japan (Katsumata 2004). Second, the four models were nominated to examine the optimum design for densely inhabited areas, which will induce more wind inside the urban domains and will improve their air quality and comfort. Third, model (II)

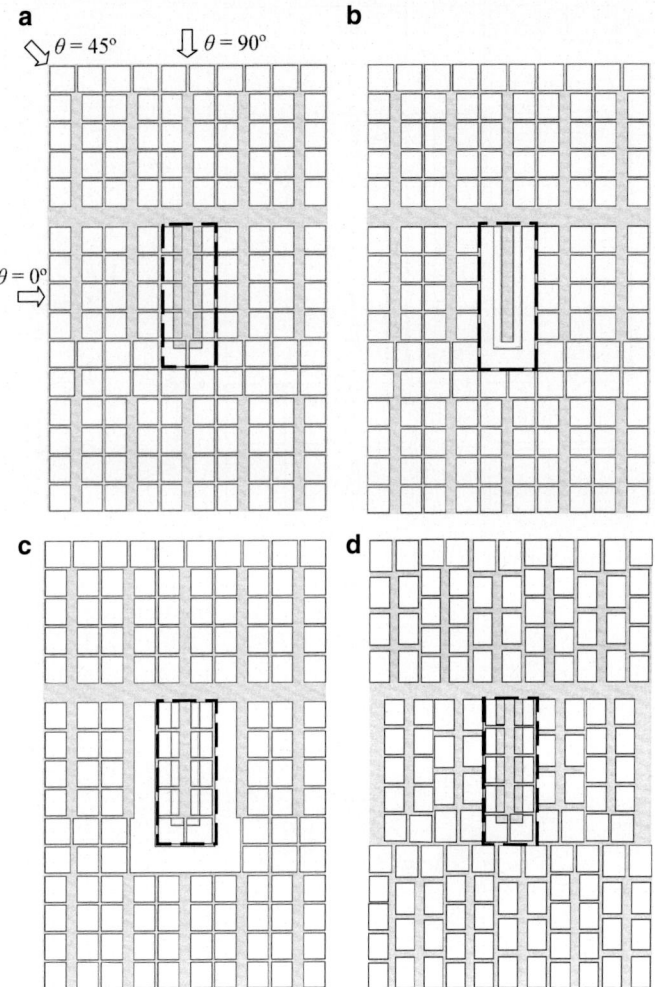

Fig. 7.10 Simplified diagrams for the four typical models of a dense urban area. (**a**) Model (I). (**b**) Model (II). (**c**) Model (III). (**d**) Model (IV)

represents an unfavorable choice for dense urban areas, due to the blocking effect of its geometry in some wind directions. The U shape of the central part of such a model decelerates the wind motion and hence traps pollutants within the pedestrian domain of the street. Fourth, model (I) represents a uniform-building array. Fifth, model (III) exhibits a combination of models (I) and (II), and as a result, it is expected to demonstrate a behavior that is intermediate between these two models. Sixth, model (IV) represents a model of what exists in reality, as the buildings in this model form a staggered array.

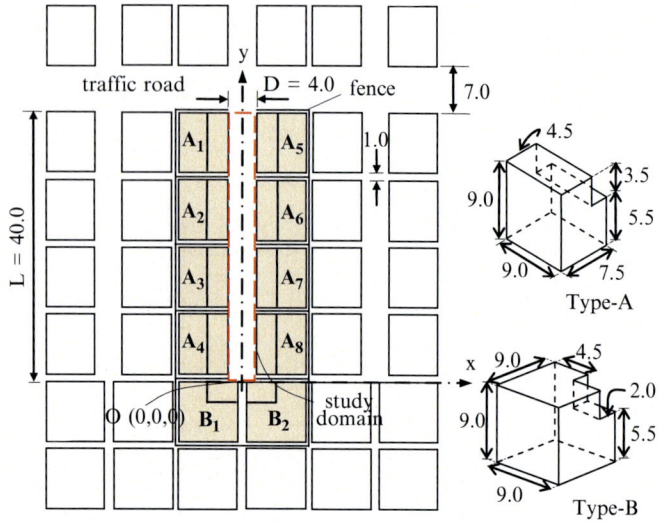

Fig. 7.11 Street characteristics in the case of model (I)

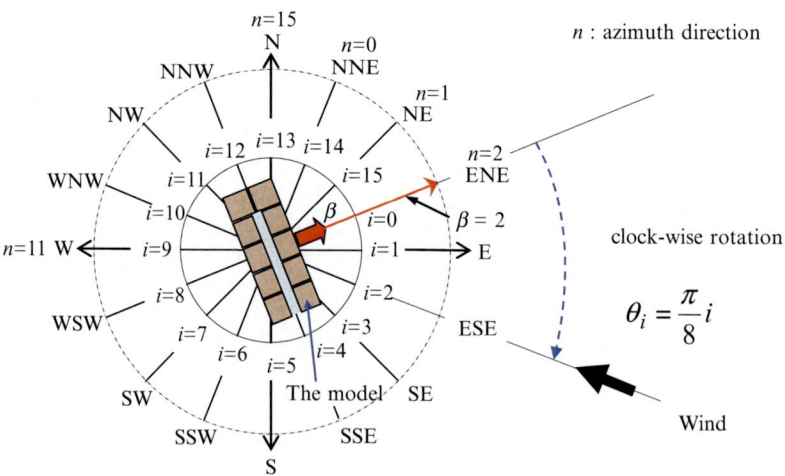

Fig. 7.12 Definitions of the dense urban areas model

In this section, nine Japanese cities were selected to conduct the exceedance probability calculations. These cities are Tokyo, Osaka, Sapporo, Niigata, Fukuoka, Nagoya, Sendai, Yokohama, and Kyoto. Figure 7.13 presents a map of Japan showing the nine cities. These cities were selected mainly because they are the most important cities in Japan, representing a significant proportion of the Japanese population and the site of many residential houses. In addition, these cities cover

Fig. 7.13 Map of Japan showing the nine assessed cities

Japan from north to south, stretching from Sapporo to Fukuoka, which means that the calculations of the exceedance probabilities for the nine selected cities will also be useful for other cities located nearby. Another important feature of the selected cities is that they have different characteristics, with some being coastal cities (both along the Sea of Japan and the Pacific Ocean) and others being inland cities. The wind regime in coastal regions is known to be completely different from those of inland areas (i.e., Kyoto).

The above reasons make the investigation of the wind characteristics of such cities very important in ensuring a reasonable level of air quality and comfort for the inhabitants of these cities.

Graphical representations of the parameter $A(a_n)$ for the nine cities are given in Fig. 7.14. The Weibull parameters for these cities were estimated by regression of the data on the mean wind velocity, which was measured every 3 h for a period of 10 years starting from 1995 until 2004 and was averaged over 10 min. The values of the Weibull parameters for the 16 azimuth directions in the nine selected Japanese cities have been previously described (Wind Engineering Institute of Japan 2005).

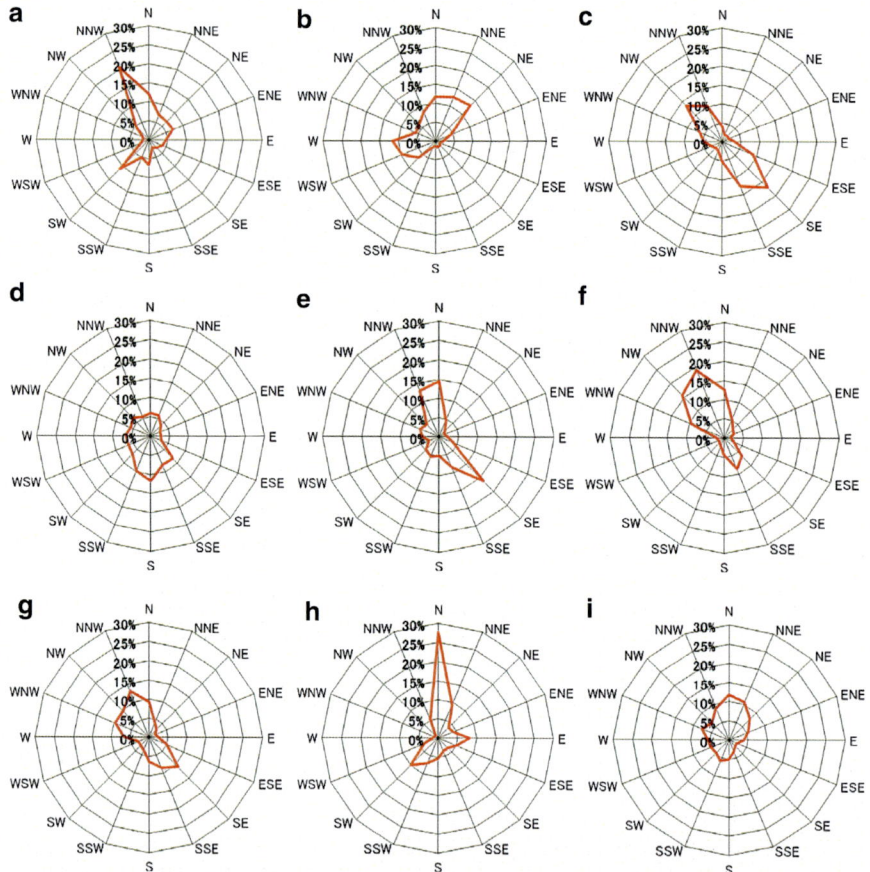

Fig. 7.14 Wind roses of the nine cities based on their mean wind velocity (averaged over 10 min).
(**a**) Tokyo (74.5). (**b**) Osaka (22.9). (**c**) Sapporo (31.1). (**d**) Niigata (15.0). (**e**) Fukuoka (24.4).
(**f**) Nagoya (17.9). (**g**) Sendai (52.0). (**h**) Yokohama (19.5). (**i**) Kyoto (16.1) (The numbers in
brackets refer to the observation height in meters)

7.5.2 Numerical Simulations

7.5.2.1 Calculation Settings

Numerical simulations were performed using the CFD code Star-CD and were based
on a finite-volume discretization method. A steady state analysis was adopted, and
the monotone advection and reconstruction scheme (MARS) was applied to the
convective term. The standard k-ε model was used to simulate the turbulence effects.
The pressure/velocity linkage was solved via the SIMPLE algorithm (Patankar
1980). As the area of study contains many geometric configurations, it was preferred

Fig. 7.15 Schematic
representation of a grid
system to simulate the array
of model (IV). (**a**) Grid
system of the whole array.
(**b**) Grid system of the study
domain

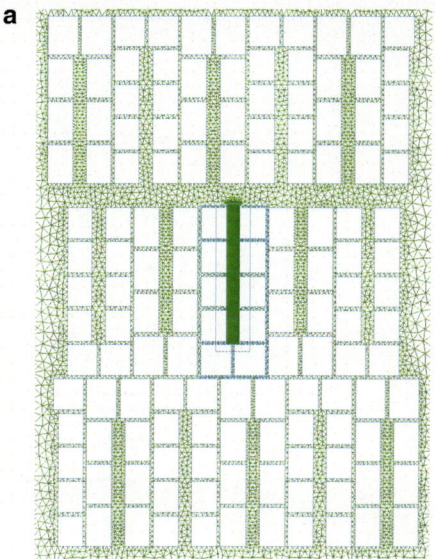

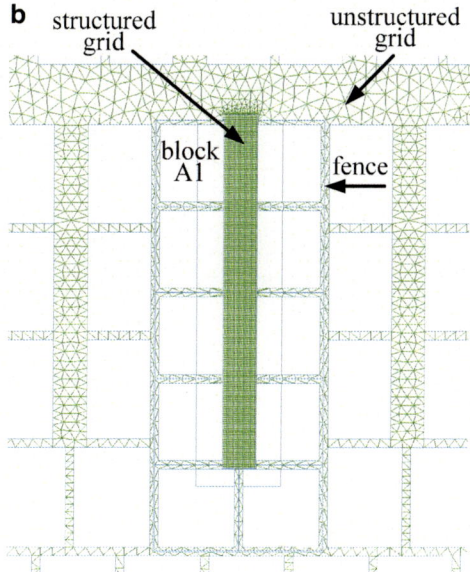

to use a system of unstructured grids because this type of mesh is suitable for CFD
simulations of complex urban areas, as demonstrated by Huang et al. (2005).
However, a structured grid system was used in the study domain because of the
expectation of greater simulation accuracy. Figure 7.15 shows a schematic representation of the grid development for the building array of model (IV), together with
the grid characteristics of the street domain. The detailed analysis conditions are the
same as in application 1, which are shown in Table 7.1.

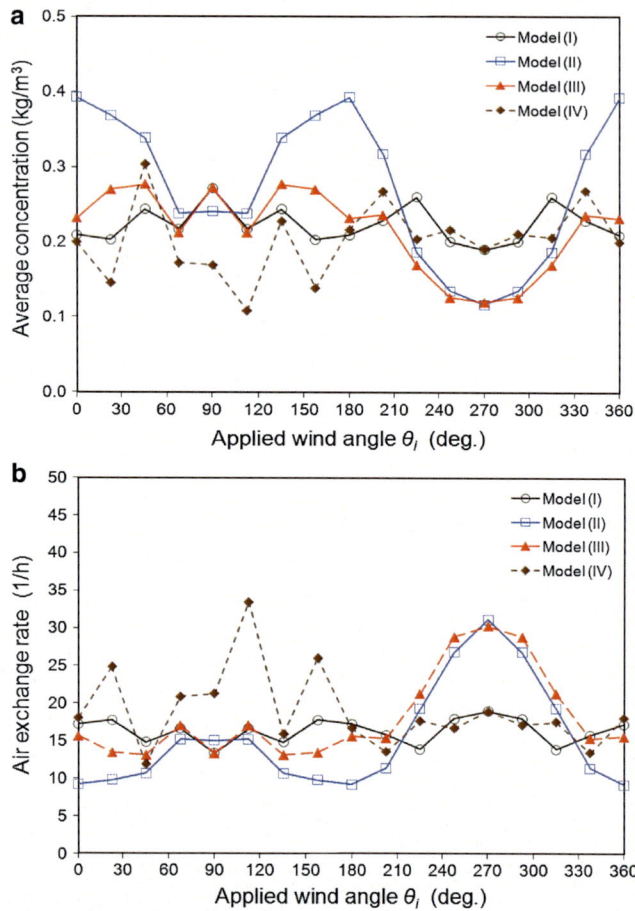

Fig. 7.16 Air quality parameters for the four models versus the applied wind directions. (**a**) Average domain concentration. (**b**) Air change rate within the domain

7.5.2.2 Calculation Results of Air Quality Within the Study Domain

Figure 7.16 shows the air quality parameters of the average pollutant concentration and the air change rate within the pedestrian domain of the street for the four building patterns and the applied wind directions. The results demonstrate a significant dependence on the wind direction and on the building array geometry. In the range from 0° to 180°, models (I) and (IV) demonstrate good ventilation performance as compared to the other models. The behaviors of such models for this wind direction range reflect the high efficiency of removal of the pollutants by the wind, which is attributed to the presence of narrow gaps between street buildings. The narrow gaps induce more wind into the street domain, which improves the ventilation process. The effects of these gaps also appear clearly in model (III), which

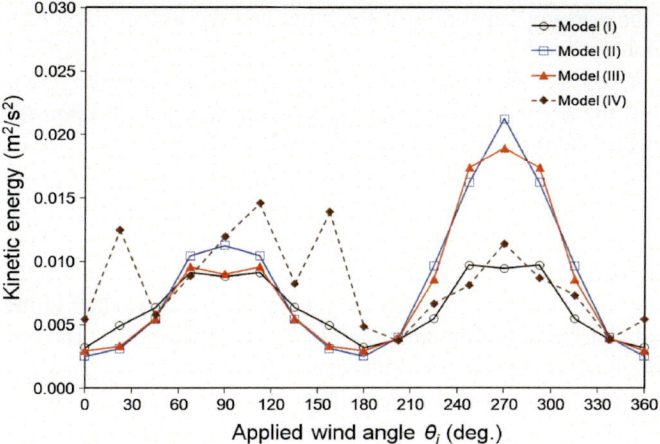

Fig. 7.17 Average wind kinetic energy within the study domain for the applied wind directions

demonstrates better performance than model (II), which has the lowest ventilation performance of the four models within this range of wind directions. It is thought that the geometry of the solid U shape in model (II) and the outer U shape of model (III) are the main reasons for these trends within this range of wind directions. The U-shaped geometry prevents the removal of pollutants by the applied wind in specific wind directions, which leads to an increase in the returning flux of these pollutants to the domain. The pollutants then accumulate, and the concentration increases.

In the range from $\theta = 180$–360, models (II) and (III) clearly demonstrate the highest ventilation performances of the four models. The effect of the wind direction within such a range can be understood by realizing that most of the wind that enters the street domain comes through the shear layer at the building roof level, while a low percentage of the wind enters the domain through the traffic road side and the narrow gaps between the street buildings. The wind flow through the narrow gaps creates lateral flows, which appear clearly in models (I) and (IV) than in model (III) and which disappear completely in the case of model (II). The interaction between these flows and the main flow coming across the shear layer creates a number of vertices within the street, which affect the purging capability of the main flow.

7.5.2.3 Calculation Results of the Wind Kinetic Energy Within the Study Domain

The average wind kinetic energy within the study domain is presented in Fig. 7.17. Clearly, the trends shown are similar to those of the air change rate presented in Fig. 7.16b. This can be attributed to the fact that the greater the wind kinetic energy, the greater the purging capability of the wind and the higher the air change rate within the domain (and vice versa).

Similarly to the air change rate trends, the trends for the average wind kinetic energy within the domain demonstrate significant dependence upon the incident wind direction. Within the range $\theta = 0{-}180°$, model (IV) nearly always demonstrates the highest values as compared with the other models, while within the range $\theta = 180{-}360°$, models (II) and (III) exhibit the highest values for the wind kinetic energy, especially at $\theta = 270°$.

An important observation from Figs. 7.14 and 7.15 is the oscillation that appears in the case of model (IV). These oscillations are attributed to the effect of the staggered distribution of the building blocks in such a model. The nature of the staggered distribution of the blocks influences the amount of wind entering the street for specific wind directions and hence affects the ventilation performance and the wind potential within the domain.

From the calculation results presented in Figs. 7.14 and 7.15, it is obvious that there is considerable variation in the domain's wind potential for the same building pattern in different wind directions. To attain a reasonable assessment for the four building patterns, further analysis of the wind environment within the study domain in these models is needed. Thus, here we adopt the use of the exceedance probability criterion to carry out such an assessment.

7.5.3   Exceedance Probability Analysis

7.5.3.1   Investigation of Model Location by Exceedance Probability Analysis

The results of calculating the exceedance probabilities within the study domain for model (I) are presented in Fig. 7.18 in terms of the air change rate that is estimated for different arrays of directions. For each city, there are 16 curves representing the 16 directions considered for the building arrays. These curves represent the exceedance probability of the air change rates for each direction of the array. The figure shows that at higher exceedance probabilities, the difference in the EP values from one direction to another is very small, while at low probability values, such differences are considerable. For example, in the case of Tokyo, at a probability value of 80%, the LACR is approximately 30 1/h in almost all directions of the array, with no considerable differences between these directions. However, at an exceedance probability of 20%, the LACR ranges from 70 1/h in the WSW direction to 77 1/h in the ENE direction. Additionally, the figure demonstrates that for a specific air change rate and a specific array direction, the probability values vary from one city to another due to the variation of the wind characteristics between these cities. In addition, the figure demonstrates the low dependence of model (I) on the array direction for almost all cities. This lack of dependence is shown clearly in the convergence between the probability curves in the 16

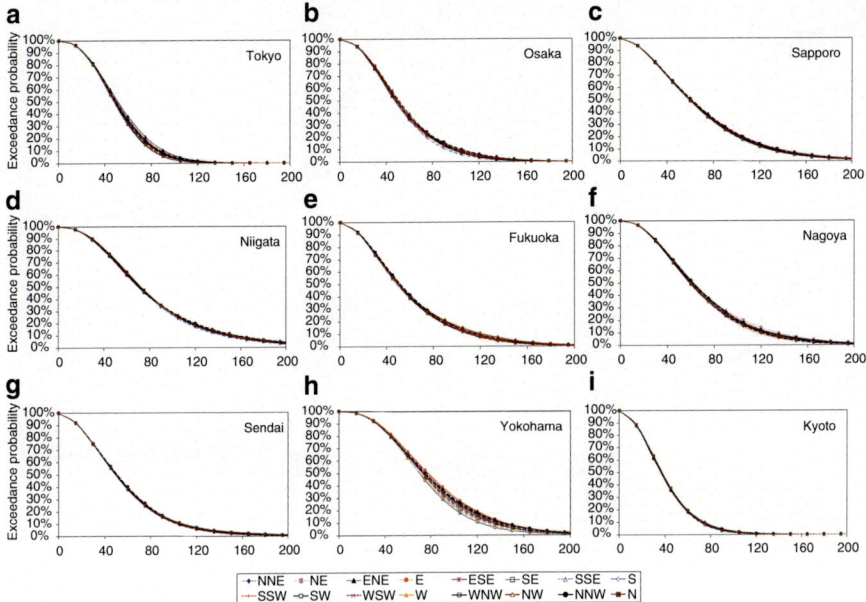

Fig. 7.18 Exceedance probability trends in terms of the local air change rate (1/h) in the case of Model (I) for the nine cities. (**a**) Tokyo. (**b**) Osaka. (**c**) Sapporo. (**d**) Niigata. (**e**) Fukuoka. (**f**) Nagoya. (**g**) Sendai. (**h**) Yokohama. (**i**) Kyoto

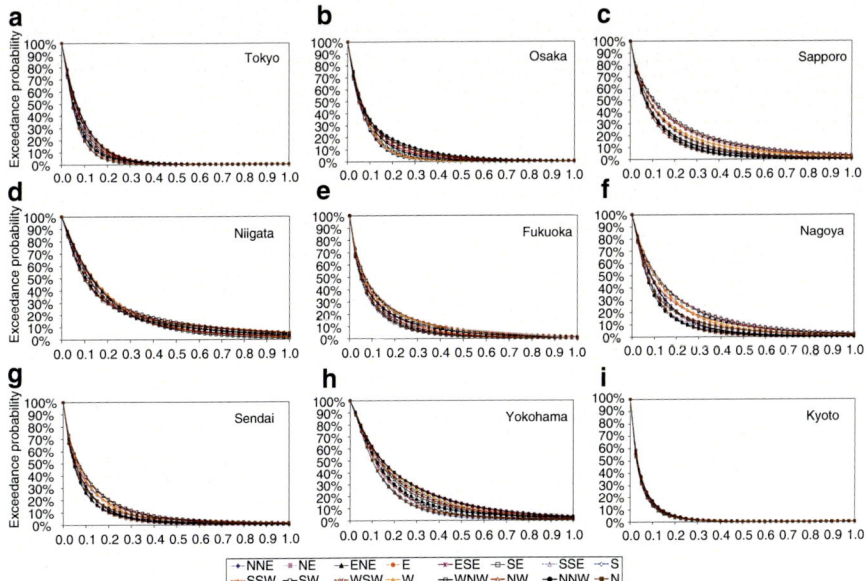

Fig. 7.19 Exceedance probability trends in terms of kinetic energy (m^2/s^2) in model (I) for the nine cities. (**a**) Tokyo. (**b**) Osaka. (**c**) Sapporo. (**d**) Niigata. (**e**) Fukuoka. (**f**) Nagoya. (**g**) Sendai. (**h**) Yokohama. (**i**) Kyoto

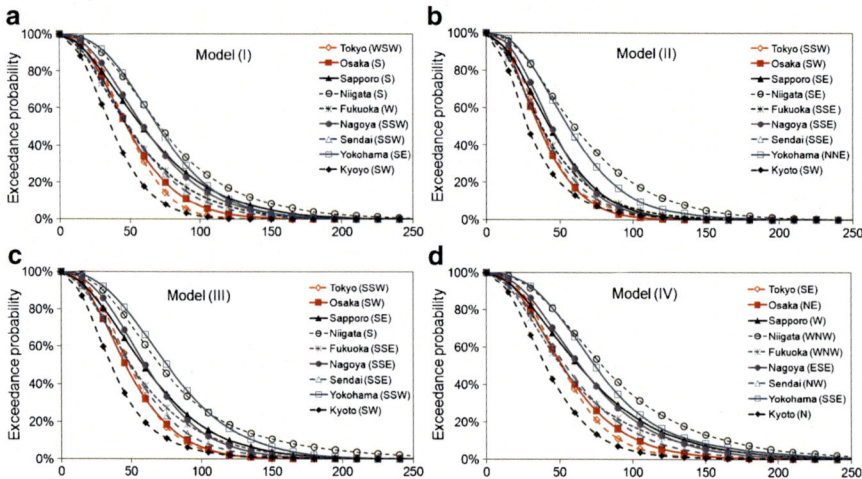

Fig. 7.20 Minimum exceedance probabilities of the four building patterns for the nine cities based on the local air change rate (1/h). (**a**) Model (I). (**b**) Model (II). (**c**) Model (III). (**d**) Model (IV)

directions of the array. Therefore, the change in the air change rate in response to changes in the array direction is very small in these kinds of densely built-up areas. Another important point to note from this figure is that model (I) has low probability values at higher ventilation rates in Tokyo and Kyoto. This can be attributed to the wind characteristics of these two cities, reflected by the values of the parameters C (a_n) and $K(a_n)$.

Figure 7.19 shows the exceedance probability of model (I), estimated by means of the average wind kinetic energy within the study domain. Similar to the probability trends shown in Fig. 7.18, low probabilities are detected at high kinetic energy values, especially in Tokyo and Kyoto, while higher probabilities are found at high kinetic energy values in Sapporo, Niigata, and Yokohama. Unlike the probability estimated using LACR, the exceedance probability estimated using wind kinetic energy shows little divergence between the probability values of the 16 wind directions.

The results of Fig. 7.16 demonstrate that it is difficult to evaluate the results of the exceedance probabilities for all 16 directions due to the high density of the curves. Therefore, it was preferable to compare the ventilation performance of the four building models based on the values of both the air change rate and the wind kinetic energy, estimated at the minimum probabilities. The minimum exceedance probabilities reflect poor air quality conditions and low comfort conditions for inhabitants of these areas, which is a very important subject of investigation.

Table 7.3 Air change rates for exceedance probability values of 80% and 20% (Unit: 1/h)

City	Model (I)		Model (II)		Model (III)		Model (IV)	
	$P = 80\%$	$P = 20\%$	$P = 80\%$	$P = 20\%$	$P = 80\%$	$P = 20\%$	$P = 80\%$	$P = 20\%$
Tokyo	29.9	69.0	22.7	57.2	29.9	71.5	31.8	76.0
Osaka	27.8	74.5	20.3	56.7	26.3	73.0	29.4	83.4
Sapporo	29.9	99.2	22.6	69.4	30.5	96.0	32.4	107.2
Niigata	42.2	113.0	30.2	101.8	41.6	115.5	45.7	128.8
Fukuoka	25.0	85.0	19.0	65.9	25.4	87.0	26.2	93.8
Nagoya	34.1	94.2	26.0	67.3	35.5	91.1	36.1	102.6
Sendai	26.2	81.3	19.8	60.8	26.3	79.2	28.2	89.9
Yokohama	44.2	101.7	33.1	85.3	45.9	113.0	45.9	112.0
Kyoto	20.2	57.7	14.9	49.7	19.2	59.4	21.6	65.3

7.5.3.2 Minimum Exceedance Probabilities Based on the Local Air Change Rate

Figure 7.20 shows the minimum exceedance probabilities of the four building patterns for the nine cities considered based on the local air change rate and the array direction at which such probabilities were calculated.

Indeed, there is a relationship between the wind directions expressed by the wind rose data (shown in Fig. 7.14) and the array direction at which the minimum probability is calculated. However, the incident wind direction is not the only parameter controlling the direction at which the lowest probability exists. For any direction of the building array, the calculated probability is the summation of the probabilities of the 16 wind directions. The combination of the wind rose data and the air exchange rate distributions shown in Fig. 7.16b determines the array direction of the minimum exceedance probability. As an example, consider the case of model (I) in Tokyo when the array direction is WSW and the wind direction is NNW. In this case, the wind direction comes in at an angle of 90° to the building array direction. From Fig. 7.16b, the air change rate for the case of model (I) has its minimum value at $\theta = 90$. Simultaneously, the NNW direction has the maximum relative frequency of occurrence in Tokyo (see Fig. 7.14). The combination of the array direction for the minimum air change rate and the wind direction for the maximum occurrence frequency results in the WSW direction being the direction of minimum probability for such a case.

From the lowest probability curves shown in Fig. 7.20, it is clear that the effective range of such probabilities lies between $P = 20\%$ and 80%. For $P < 20\%$, the air change rate is very high, which is not common, while for $P > 80\%$, the air exchange rate is quite low, and the air quality conditions are poor. Thus, it is important to analyze the LACR values within this range. To do so, one must focus on the air change rates at the upper and the lower limits of such a range. Table 7.3 presents the values of the air change rate for the four building patterns in the nine cities at $P = 80\%$ and 20%. Additionally, a graphical representation of these values is given in Fig. 7.21.

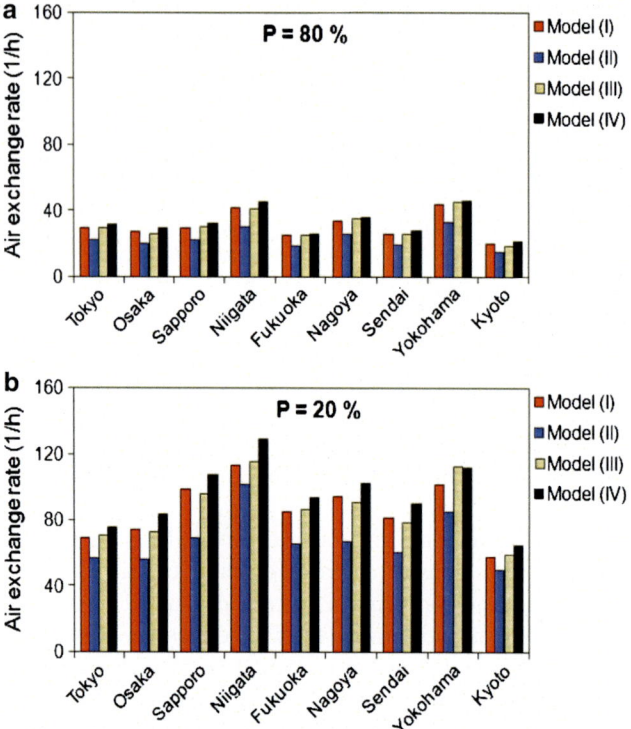

Fig. 7.21 Air change rates (1/h) at probability values of $P = 80\%$ and 20% for the nine cities. (**a**) Air change rate for $P = 80\%$. (**b**) Air change rate for $P = 20\%$

Here, $P = 80\%$ corresponds to an approximation of an exceedance probability of 6/7, as mentioned above in Sect. 7.3.6, and $P = 20\%$ corresponds to an exceedance probability of 1/7. Both are the same in practice, and for simplicity, this section used $P = 80\%$ and 20% to make the illustration.

As can be seen from Table 7.3 and Fig. 7.20, the lowest air change rate values within the study domain for the four models are found in Kyoto at both $P = 80\%$ and 20%, while the highest values are between Niigata and Yokohama. This result demonstrates that the wind conditions in Yokohama and Niigata are better than those of the other seven cities, even in weak wind conditions. It is also clear that the difference between the air change rates of the nine cities at $P = 80\%$ is not very high, while at $P = 20\%$, the difference is remarkable.

In addition to the above, Table 7.3 demonstrates that model (II) has the lowest LACR values of the four, while models (II), (III), and (IV) show nearly the same values. As the minimum standard values of the air change rate within these cities are not determined precisely, it is difficult to state that model (II) is unacceptable. Referring to Fig. 7.20, at a low LACR of up to 25 (1/h), the difference between model (II) and the other models is not particularly significant. Thus, the performance of the four building models at low air exchange rates is nearly identical.

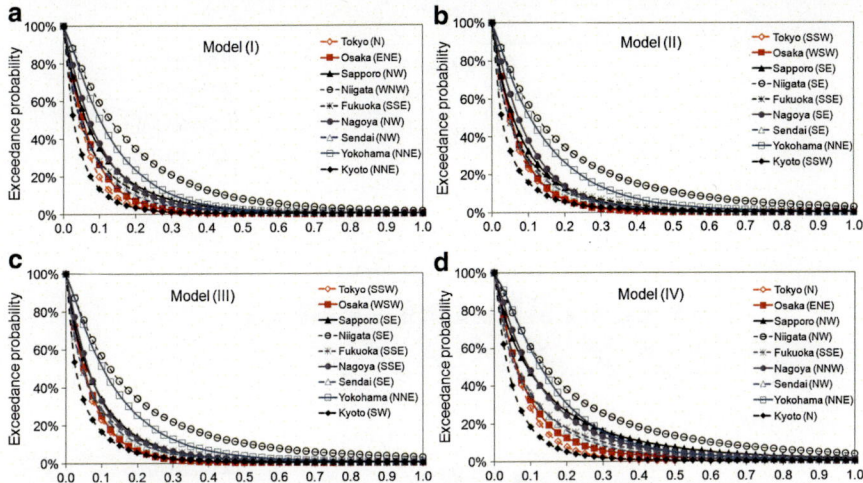

Fig. 7.22 Minimum exceedance probabilities of the four building patterns for the nine cities based on kinetic energy (m^2/s^2). (**a**) Model (I). (**b**) Model (II). (**c**) Model (III). (**d**) Model (IV)

Table 7.4 Kinetic energy for exceedance probability values of 80% and 20% (Unit: m^2/s^2)

City	Model (I)		Model (II)		Model (III)		Model (IV)	
	$P = 80\%$	$P = 20\%$	$P = 80\%$	$P = 20\%$	$P = 80\%$	$P = 20\%$	$P = 80\%$	$P = 20\%$
Tokyo	0.0180	0.099	0.0178	0.111	0.0185	0.111	0.0235	0.125
Osaka	0.0175	0.120	0.0170	0.118	0.0170	0.117	0.0210	0.148
Sapporo	0.0175	0.167	0.0165	0.153	0.0170	0.159	0.0172	0.260
Niigata	0.0440	0.309	0.0400	0.330	0.0395	0.321	0.0470	0.375
Fukuoka	0.0125	0.138	0.0120	0.142	0.0120	0.143	0.0155	0.182
Nagoya	0.0280	0.165	0.0250	0.164	0.0250	0.151	0.0320	0.232
Sendai	0.0135	0.127	0.0130	0.130	0.0130	0.130	0.0180	0.175
Yokohama	0.0390	0.220	0.0370	0.242	0.0380	0.234	0.0500	0.265
Kyoto	0.0080	0.076	0.0074	0.083	0.0080	0.086	0.0112	0.095

7.5.3.3 Minimum Exceedance Probabilities Based on Kinetic Energy

The minimum exceedance probabilities of the four building patterns for the nine cities, based on the average wind kinetic energy within the domain, are shown in Fig. 7.22, together with the direction at which these probabilities were calculated.

As mentioned in the previous section, the investigation of the kinetic energy values at $P = 80\%$ and $P = 20\%$ is very important because the operating range is between these two limits. Table 7.4 presents the kinetic energy values of the wind for the four building patterns, based on the wind conditions of the nine cities, at $P = 80\%$ and 20%. These values are also presented graphically in Fig. 7.23.

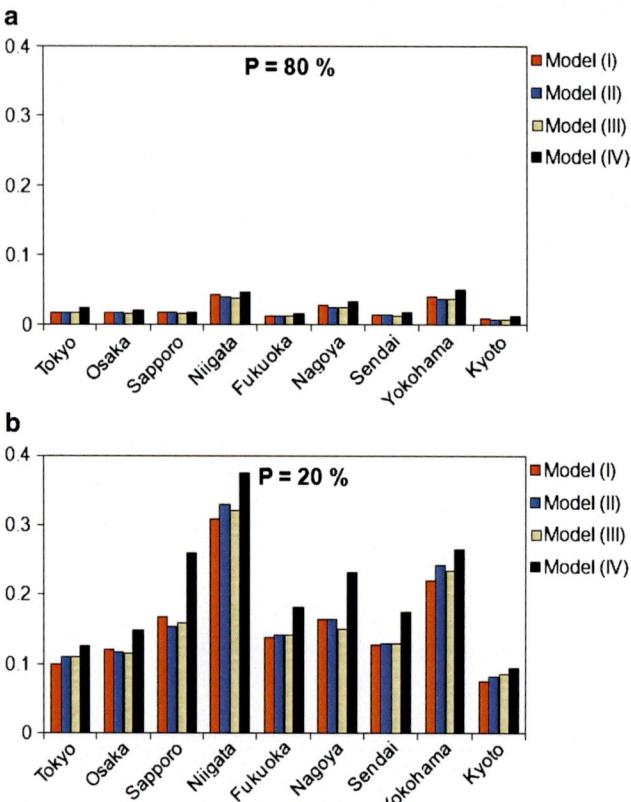

Fig. 7.23 Kinetic energy (m^2/s^2) at probability values of $P = 80\%$ and 20% for the nine cities. (**a**) Kinetic energy for $P = 80\%$. (**b**) Kinetic energy for $P = 20\%$

The results demonstrate the considerable differences between the kinetic energy values estimated at $P = 80\%$ and at $P = 20\%$ in this case (based on the kinetic energy) compared with those of the previous case (based on the air change rate). These differences can be attributed to the slopes of the probability curves that are estimated in terms of the kinetic energy, which are greater than the probability curves estimated in terms of the air change rate.

From the values given in Table 7.4 and Fig. 7.23, the lowest kinetic energy values within the study domain are found in Kyoto for both $P = 80\%$ and $P = 20\%$. In contrast, the highest kinetic energy values are for Niigata and Yokohama. This result confirms that the wind conditions in Kyoto are the worst of these nine cities, while those of Niigata and Yokohama are the best, even in weak wind conditions. Another important point to note is that the differences between the kinetic energy values of the nine cities estimated at $P = 20\%$ is higher than the differences between the values estimated at $P = 80\%$.

Table 7.5 Recommended directions for the four building arrays in the nine cities (Directions of maximum exceedance probabilities)

City	Model (I) ACR	Model (I) KE	Model (II) ACR	Model (II) KE	Model (III) ACR	Model (III) KE	Model (IV) ACR	Model (IV) KE
Tokyo	ENE	E	E	E	E	E	SW	NW
Osaka	ESE	NNW	E	NNW	E	NNW	WSW	SE
Sapporo	NE	NE	SW	SW	SW	SW	NNW	NE
Niigata	NNW	SW	W	NE	W	NNW	ENE	SW
Fukuoka	E	WSW	ENE	ENE	ENE	ENE	WSW	WSW
Nagoya	NE	NE	NE	ENE	NE	ENE	SW	SW
Sendai	NNW	WSW	NE	NE	NE	NE	SW	SW
Yokohama	E	ESE	ESE	ESE	E	ESE	WSW	E
Kyoto	S	N	ESE	ESE	ESE	ESE	W	W

7.5.3.4   Recommendations for the Best Directions of the Four Building Arrays

In this subsection, recommendations for the best directions of the four building arrays, at which the maximum exceedance probability occurs, are introduced. These directions are those of the maximum exceedance probabilities that were calculated based on both the air exchange rate and the wind kinetic energy. The choice of array directions depends upon the requirements set by the designers. If the array direction is selected based on the air change rate, then the air quality is the preferred parameter of the designer. If the array direction is selected based on the wind kinetic energy, then comfort is the desired parameter. However, the application of such recommendations basically depends upon the potential to connect existing roads within the city in question with those passing through the building array. Table 7.5 presents the directions of the building arrays for the nine cities based on the maximum exceedance probabilities. The authors strongly recommend the application of these directions wherever possible. However, it is important to mention that the directions of the building arrays that were estimated based on the wind kinetic energy are recommended only for weak wind conditions.

7.6   Conclusions

In this chapter, the concept of exceedance probability was introduced and employed as the assessment approach in the investigation of the local wind environment at the pedestrian level. Two new criteria, the local air change rate-based exceedance probability criterion and the local kinetic energy-based exceedance probability criterion, were proposed for the exceedance probability analysis to comprehensively take into account the uncertainties associated with the influence of the climate on the wind environment.

Compared with the traditional velocity-based criteria, these two criteria are essentially domain-based criteria, which can enable an evaluation of the reliability and acceptability of the local wind environment in terms of wind-driven ventilation efficiency and thermal comfort. In the construction of the proposed criteria, the local air change rate and the local kinetic energy were used as the indices with regard to the ventilation efficiency and the wind intensity, respectively, to take into account the wind characteristics involved in dispersing contaminants and improving thermal comfort.

Two application examples were given for detailed illustrations. In the first example, an idealized street canyon was defined and used to show the application of the proposed criteria in evaluating the local wind environment at ground level. The influences of two factors, the orientation and width of the void model, were investigated in detail by using both criteria based on CFD simulation. The simulation results demonstrate that both factors have great influences on the local ventilation performance, which also demonstrates the practicability and effectiveness of the proposed criteria to help achieve a good wind environmental design. The exceedance probability analysis method also seems to be an effective tool for evaluating urban ventilation. It can estimate the optimum direction for building arrays that provide a good wind environment for the inhabitants of such areas by improving the air quality and human comfort conditions.

In the second example, four typical models of densely built-up areas located in nine cities in Japan were studied and numerically simulated to examine the effects of the geometry of these building patterns on the wind flow characteristics within the pedestrian domain. The results indicate that exceedance probabilities strongly depend upon the geometries of the four building arrays as well as on the wind conditions of the construction site. Dense building arrays with some narrow gaps can be set in any direction within the construction site, and the presence of narrow gaps leads to a decrease in the differences between the air exchange rates for variable wind directions. Based on the minimum probability values in terms of the air change rate and wind kinetic energy, the wind conditions of Yokohama and Niigata seem to be better than other cities. This conclusion is drawn from the weak wind point of view. Finally, we present recommendations for the best directions of the four building arrays, which enhance the wind and attain the maximum ventilation performance and the best comfort conditions within the pedestrian domain. However, the application of these recommendations depends upon the potential to connect existing roads within the city in question to those passing through the building array. Also, the directions of the building arrays that were estimated based on the kinetic energy are recommended only for weak wind conditions.

In the future, this research will be focused on applications for real-world problems, with the effects of buoyancy and vehicle movement taken into account.

Acknowledgments We would like to express our thanks to Dr. Takao Sawachi, Dr. Katsumata Wataru, and Dr. Ishii Norimitsu of the National Institute for Land and Infrastructure Management for giving valuable advice in fulfilling this research.

References

CD Adapco Group (2004) STAR-CD methodology, (Version 3.15). Computational Dynamics Limited, Yokohama

Huang H, Ooka R, Kato S (2005) Urban thermal environment measurements and numerical simulation of an actual complex urban area covering a large district heating and cooling system in summer. Atmos Environ 39:6362–6375

Huang H, Kato S, Ooka R (2006) CFD analysis of ventilation efficiency around an elevated highway using visitation frequency and purging flow rate. Wind Struct 9:297–313

Hunt JCR, Poulton EC, Mumford JC (1976) The effects of wind on people; new criteria based on wind tunnel experiments. Build Environ 11:15–28

Hunter LJ, Johnson GT, Watson ID (1992) An investigation of three-dimensional characteristics of flow regimes within the urban canyon. Atmos Environ 26:425–432

Ishida Y, Kato S, Huang H, Ooka R (2005) Study on wind environment in urban blocks by CFD analysis. In: Proceedings of the 6th Asia-Pacific conference on wind engineering, Seoul, pp 397–541

Ito K, Kato S, Murakami S (2000) Study of visitation frequency and purging flow rate based on averaged contaminant distribution—study on evaluation of ventilation effectiveness of occupied space in room. J Archit Plann Environ Eng 529:31–37 (in Japanese)

Japan Association for Wind Engineering (2005) The basic knowledge about building wind. Kajima Publishing Co., Ltd, Tokyo (in Japanese)

Katsumata W (2004) Strategies for improving the residential environment of existing suburban small-lot residential areas through harmoniously controlling the rebuilding of houses. PhD thesis, University of Tokyo, Japan (in Japanese)

Murakami S, Iwasa Y, Morikawa Y (1986) Study on acceptable criteria for assessing wind environment at ground level based on residents' diaries. J Wind Eng Ind Aerodyn 24:1–18

Murakami S, Mochida A, Hayashi Y (1988) Modification of production terms in k-ε Model to remove overestimate of k value around windward corner. In: Proceedings of the 10th wind engineering symposium, Tokyo, Japan (in Japanese), pp 199–204

Ohba M, Kobayashi N, Murakami S (1988) Study on the assessment of environmental wind conditions at ground level in a built-up area—based on long-term measurements using portable 3-cup anemometers. J Wind Eng Ind Aerodyn 28:129–138

Oke TR (1988) Street design and urban canopy layer climate. Energy Build 11:103–113

Patankar S (1980) Numerical heat transfer and fluid flow. Hemisphere Publishing Corporation, New York

Peng SH, Davidson L (1997) Towards the determination of regional purging flow rate. Build Environ 32:513–525

Penwarden AD, Wise AFE (1975) Wind environment around buildings. Building Research Establishment Report. Garston, UK

Pietrzyk K, Hagentoft CE (2008a) Reliability analysis in building physics design. Build Environ 43:558–568

Pietrzyk K, Hagentoft CE (2008b) Probabilistic analysis of air infiltration in low-rise buildings. Build Environ 43:537–549

Sandberg M, Sj berg M (1983) The use of moments for assessing air quality in ventilated rooms. Build Environ 18:181–197

Sini JF, Anquetin S, Mestayer PG (1996) Pollutant dispersion and thermal effects in urban street canyons. Atmos Environ 30:2659–2677

Wind Engineering Institute of Japan (2005) Fundamentals of wind around buildings. Kajama Publisher, Tokyo (in Japanese)

Index

A
Acceptable frequency, 139, 140, 160
Acceptable wind speed level, 139
Adaptive model, 45
Adaptive thermal comfort, 34
Air change rate, 156–159, 161–164, 167, 169–175, 182–192
Air sampling, 69, 73–75, 78, 79, 81, 85
Air ventilation assessment system (AVAS), 143–144, 148
Architectural institute of Japan, 13, 138
ASHRAE scale, 39, 40, 43, 44, 54, 55
AVAS. *See* Air ventilation assessment system (AVAS)

B
Bangkok, 44, 47, 60, 61, 68–91
Bedford scale, 35, 51
BEE. *See* Building environmental efficiency (BEE)
BEE-HI. *See* Building environmental efficiency for heat island relaxation (BEE-HI)
Benzo[a]pyrene (BaP), 61–63, 65–68, 78, 80–85, 87, 89–91
Briggs equation, 127, 129, 130
Building coverage ratio, 137–139
Building density, 2, 5, 154
Building environmental efficiency (BEE), 141, 142
Building environmental efficiency for heat island relaxation (BEE-HI), 142
Building Standard Law, 137–139, 148
Built-up urban area, 152, 158, 159
Bulk control system, 137–139

C
Carcinogenicity, 66, 68
CASBEE, 141–143
CASBEE-HI, 141–143, 148
Ceiling fan, 52, 53
Climate, 12–17, 34, 36–38, 41, 42, 45, 50–52, 54, 55, 68, 79, 191
Coastal area, 15, 26, 27, 29
Comfort zone, 54, 55
Commercial zone, 137, 138
Cool spots, 145, 146

D
Daily maximum gust wind speed, 139, 140, 160
Diurnal variation, 24, 29, 74–79, 89

E
Effect of the air movement, 51
Exceedance probability, 153, 160–167, 172–176, 178, 184–192
Exponential distribution, 118, 127
Exposure, 4, 5, 59–92, 98, 104

F
Fick's law, 99–103, 106
Field measurement, 60, 68–80, 90, 111
Floor area ratio, 2, 137, 138
Free-running mode, 48, 49

G
Gaussian distribution, 107–109, 117
Geospatial Information Authority of Japan, 19, 28

S. Kato and K. Hiyama (eds.), *Ventilating Cities: Air-flow Criteria for Healthy and Comfortable Urban Living*, Springer Geography, DOI 10.1007/978-94-007-2771-7, © Springer Science+Business Media B.V. 2012

ISBN 978-9400727700 -0

Printed in Japan

C0001588

9 789400 727700